VALOR TALKS

VALOR TALKS

Eyewitness Accounts of World War II's Greatest Battles

KEVIN DENNEHY

GTCI Press
Denver, Colorado

Valor Talks
First Edition

Visit our website: www.militaryhistorytraveler.com.
Photographs and other images were provided by the interviewed veterans or their families.

Published by GTCI Press

ISBN-13: 978-0-9971105-2-4

This book is dedicated to those who never had the chance to tell their stories...the 407,316 Americans,

from all military branches,

who died while in service during World War II.

Henry "Hank" Cornellisson, U.S. Army Air Corps, spent 1,212 days as a Japanese POW during World War II. Hank fought on Corregidor and Baatan and was captured in 1942. He survived a Japanese "hell ship" that took him to a camp to work as a slave laborer. He was liberated in September 1945 and stayed in the military. He returned to the United States, married Ruth Elizabeth Jordan and had three sons.

CONTENTS

INTRODUCTION

From Pearl Harbor to Tokyo Bay, more than 16 million men and women served in World War II. Of the more than 16 million, as of 2023, there are only about 119,550 left, with 180 dying each day, according to the U.S. Department of Veterans Affairs.

The memories of a long-ago war fade each day. These veterans experienced the sights and sounds of the world's most devastating war that still have ramifications today.

Soon, there will not be an eye-witness account of what happened when Pearl Harbor was bombed, the D-Day landings, when Nazi concentration camps were liberated, and the atomic bombs dropped.

This is where I hope this book of interviews serves as a reminder that real people experienced the triumphs and horrors of a war that was not as long ago as it seems—at least while there are still men and women alive to tell their stories. It was not only gratifying talking with these veterans, but immensely humbling.

Sadly, at this writing, most of the interviewed veterans have passed away. However, their memory lives on through these interviews—from a concentration camp liberator, a survivor of the USS *Indianapolis*, a crewman who dropped the atomic bomb on Nagasaki, two Easy Company Band of Brothers members and many others.

We also include two interviews with veterans who support the memory of our World War II veterans. One is an interview with the superintendent of an American Battle Monuments Commission cemetery. The other is with the former executive director of the board that oversees the National Museum of the Pacific War in Texas.

A goal of this book is to allow you, the reader, to not only read these tales of World War II valor, but be able to conduct your own interviews with the remaining survivors of an epic war. Whether they be family, friends, or even strangers, they each have unique stories that need to be told—before it's too late.

Only one percent of Americans serve in the U.S. military. The majority of the other 99 percent don't regularly come into personal contact with service members. They really don't know the people who defended their freedoms and way of life. This is your chance to let them know by telling their stories.

Our appendix section has tips on how to interview/talk to a veteran that may prove valuable for first-time interviewers. We also have examples of questions to ask and tips to make a veteran comfortable during the interview session. In the actual interviews, written in bold, we've placed background information about the unit or the battle where the veteran fought.

Our website, www.militaryhistorytraveler.com, offers you other examples of veteran interviews to assist you to prepare your own.

I look forward to your contributions and critical input on our website, as well as Amazon reviews, as we go forward. We hope you will experience how exciting and informative it is to not only read my veterans' interviews, but in conducting your own research and interviews of this Greatest Generation.

Kevin Dennehy
Denver, Colorado
www.militaryhistorytraveler.com
September 2024

ACKNOWLEDGMENTS

I am indebted to a number of colleagues, friends and family who supported me while this book was being written. It's taken more than a decade to compile these interviews, which I could not have done without their dedicated help.

Where do I start? My friend in Normandy, France, Margaret Heesen, who assisted in interviewing Charles Shay and Benjamin Berry. Margaret has helped organize many events that commemorate our epic landings in Normandy.

The Band of Brothers interviews, conducted by Mike Van Buren, offer great insight to two Easy Company veterans. Mike uncovered real background information behind the famed HBO miniseries. I thank him for his many contacts and years of honoring our World War II veterans.

To my friend, classmate and fellow historian Rob Woodard, who took some of my questions, in addition to his own, to craft great interviews with most of the World War II aviators in this book. Rob's ability to make the veterans comfortable, even when asking uncomfortable questions about their wartime experiences, was invaluable.

There are many steps to make a book like this possible that include precise copy editing, book cover art and platform conversion. I would like to thank Jude DeLorca and Bill Andersen for their expert editorial suggestions and copy editing. Trish Shapiro, who has expertly edited and placed my books on Amazon and other platforms, was a steady hand throughout this process.

I would like to remember and thank Professor Stephen T. Powers, who wrote four other books with me. Steve, who passed away in 2020, expertly made recommendations about the content of this book and even wrote much of the background history of the battles and units of those interviewed.

Finally, I would like to thank the veterans themselves and their caregivers. They truly were, and are, the Greatest Generation. I cannot praise or thank them enough for helping me keep the memory of their service alive for future generations.

WILLIAM "WILD BILL" GUARNERE: EASY COMPANY, 506TH PARACHUTE INFANTRY, 101ST AIRBORNE DIVISION

The 101st Airborne Division (AD) became one of the most celebrated American combat divisions of World War II. And, no outfit exemplified the elite paratrooper attitude better than Easy Company, 506th Parachute Infantry, the unit immortalized in Stephen Ambrose's 1992 best-selling book *Band of Brothers* later turned into a TV mini-series (2001) by Tom Hanks, Erik Jendresen and Steven Spielberg.

The deployment of combat forces from the air was a new concept in the 1930s predicated on the rapid development of transport aircraft in the decade. The first steps by the U.S. Army had been to convert the recently activated 82nd Infantry Division into the 82nd Airborne Division. Later the 82nd was cloned to provide a trained cadre for the newly authorized 101st Airborne Division. In 1942 and 1943, the 82nd saw combat in the Mediterranean theater, while the 101st continued its training cycle in the U.S. and England. Both divisions were dropped into the Cotentin Peninsula behind Utah Beach on D-Day.

Band of Brothers' Bill Guarnere.

Betraying their infantry origin, American airborne divisions were built around three parachute infantry regiments and associated support units. One of the least appreciated of those support units for the 101st AD was the 327th Glider Infantry Regiment that actually made the 101st a four-regiment unit. Glider Infantry were trained to descend on the battlefield in WACO gliders (towed by C-47s twin-engine tugs), but in actual practice they often landed from the sea as regular infantry.

The 101st trained at Camp Toccoa Fort Benning, Fort Bragg, and Camp Mackall in the U.S. before being deployed to England. It subsequently made combat jumps into Normandy and Holland, toughed it through the epic battle at Bastogne, Belgium, and found itself occupying Hitler's Eagle Nest at the end of the war.

Before his death in 2014, we interviewed Easy Company's Staff Sgt. Bill Guarnere, who lost his right leg below the knee during the Battle of the Bulge. According to historian Stephen Ambrose, working from an oral interview, his award of an 80 percent disability on discharge in 1945 was not raised to 100 percent until many years later. One of 10 siblings, Guarnere grew up in Philadelphia, where he was living at the time of our interview.

Interviewer: When did you decide to join the Army? Why the paratroopers? Did you enjoy parachuting out of a perfectly good airplane?

Guarnere: Before the war, in Philadelphia, we never knew what was going to happen. We were thinking—are we going to get into the war? We didn't know where Pearl Harbor was. After Pearl Harbor, most kids joined the Army, Navy or Marines. I was one of them—it was the right thing to do. The paratroopers were new and exciting. I told myself, 'let me try it, if I make it, I make it.' I enjoyed the parachuting—there is nothing like it. When you talk about the best, paratroopers were the best. Other outfits think they were the best, but we were.

Interviewer: In the HBO series *Band of Brothers*, Richard Winters was revered as an example of a great leader. What are your memories of him?

Guarnere: Lt. Winters, as we knew him, took over Easy Company on D-Day. He was the best officer, but he also inherited a pretty fine company. Capt. (Herbert) Sobel trained the company. Lt. Winters was a great combat leader. Sobel could not lead in combat. The sergeants of Easy Company got rid of Sobel. The officers knew it, but couldn't do a damn thing about it.

Interviewer: Was your first company commander, Capt. Herbert Sobel, really a jerk? The book and series portrays him as one. How about the depiction of Lt. Ronald Speirs as a tough and crazy officer? How was actor Frank John Hughes as Staff Sgt. William Guarnere?

Guarnere: First of all, Sobel trained the company for war. Everyone knew he couldn't lead in combat. Speirs was in D Company, he came to E Company after Bastogne. We heard all the stories about how tough he was. He was an excellent officer.

Frank Hughes must have called me 250 times before the movie was shot to ask all kinds of questions. I told him he can't get perfection. He called me so often that I told him to sit down and remember no matter what he did in the series, we won the goddamn war—so don't worry about it. He did an excellent job. The casting was excellent. All of the stories were true. The best part of the series was the last (episode) that explained what became of the men after the war.

Interviewer: Did your training in Georgia and in England prepare you for the horrors and rigors of war? What was your favorite weapon?

Guarnere: The training was realistic. My favorite weapon was the 45-caliber Thompson submachine gun. You couldn't shoot too far with it, but it was my favorite.

Interviewer: Did you hate the Germans you faced in battle? What were your overall perceptions of the German soldiers you faced?

Guarnere: I was a wildman. When I initially went into combat, I learned my brother had been killed in Italy. You didn't want to be a German around me. The first day in combat, every German I saw, I killed him. I don't know how I made it through that day. I have been to Europe 16 times and have visited my brother's grave. He was killed at Monte Cassino. You don't realize it at the time, but you weren't alone in combat—it happened everywhere. My hatred of Germans is gone. The war is over.

I'm Italian. There were no inner conflicts about fighting a war—because it was a war of survival. Italians were once the enemy. You think about it, you are dead.

Interviewer: Normandy, Holland, Bastogne, Foy...all were tough campaigns and battles for the 101st Airborne. Is there any particular battle you feel that was the toughest or have the most memories about?

Guarnere: They were all different. Most of my memories were of D-Day. That was the biggest (operation). We linked up with the infantry coming up on the causeway from Utah Beach. In the battle for Brecourt Manor (*editor's note: Lt. Dick Winters won the Distinguished Service Cross for extraordinary heroism during the battle*), it was a fast battle, we were full of adrenaline and we were exhausted when the battle was over. The Germans did not even know we were there. After the battle, I went to the top of a hill to search for Germans. I fell over, dead asleep. Another soldier, Joe Toye, thought I was dead.

Holland was a fiasco, run by the British, a failure. Before *Operation Market Garden*, (British Field Marshal Bernard L.) Montgomery knew that there were enemy armored divisions in the vicinity. The British soldiers fought like lions, but were led by a bunch of goddamn donkeys. But the worst was Bastogne in December 1944 because of the conditions, the cold. Because of that battle, medics are still my heroes.

Interviewer: What did you do when you got home from the war? How did it change you?

Guarnere: First thing I did was get married to my girl Frannie, who was waiting for me. I had kids and went to work. I tried to forget about the war. No one talked about the war at all. I was on crutches. When people asked about them, I would tell them a shark bit my leg off. I didn't talk about the war for at least 50 years.

Interviewer: Are you planning any trips to Europe? Have you recently returned?

Guarnere: In September 2006, Babe Heffron and I are chaperoning a tour.

Interviewer: Besides *Band of Brothers*, what was the most realistic war movie or series you have seen? Anything about the *Band of Brothers* series you think could have been explained in more detail?

Guarnere: I couldn't care less about seeing a war movie. The way combat is portrayed is like a fairy tale. One seemed realistic—1949's *Battleground*, which focused on the 327th (Glider Infantry) at Bastogne. Gen. McAuliffe, from what I have heard, and I have spoken to a few generals about it, really did say 'nuts' to the Germans when they asked for us to surrender. He didn't know what else to say, Gen. Kennard said later. The 101st got all of the publicity during the Battle of the Bulge. We were surrounded, but there was fighting all over the damn place by a lot of other units.

There were a few inaccuracies in the series. There was too much cursing in the series by the soldiers. I never cursed like that around my children.

Interviewer: Do you still keep in touch with members of Easy Company? Who is the company's greatest unsung hero?

Guarnere: I still keep in touch with Babe Heffron and others. In 1947, we had our first reunion, I couldn't attend because I was still in the hospital. There were 13 Easy Company men in attendance. We asked each man to bring another man to the next reunion. Every year, I run an Easy Company reunion. Capt. (Lewis) Nixon was a good man. He used to go to the reunions.

(Four-time Purple Heart recipient) Joe Toye was the greatest unsung hero. He was from West Reading, Pa. He attended the reunions as well. I think of all of them. I think about them to keep their memory alive…the ones who have died can't keep their memory alive, so I do it.

Interviewer: Is there one thing you want people to remember about Easy Company?

Guarnere: Let people remember Easy Company, but also remember that many units and countries helped to win that war. The Navy, Air Corps and Marines. The British, Canadians, Australians, South Africans, New Zealanders, and many others. When I speak to groups, I let people know that without the U.S. Navy, we never would have won the goddamn war.

Interviewer: What is the greatest single lesson from your war experiences that carried over to civilian life?

Guarnere: There is no glory in war…it is simply hell. I can't explain it. You have to be there to experience it. It is like trying to explain freedom to kids who don't know about it. You don't know what freedom is until you lose it. When you look at the war, you survived it. Take Christmas time, in 1944, it was a different time. We had three or four years of routine 911s (in terms of daily casualties).

My brother, Henry, earned quite a few decorations as a medic. We used to correspond—and letters were tough to get. All of sudden, they stopped. (Fellow Easy Company soldier) Johnny Martin told me he had been killed. My older brother, Ernest, was in the Merchant Marine and was torpedoed and was listed for six months as missing in action. You are young and fighting a war. You never give a thought what your parents went through—three sons in the service—all in combat zones.

Interviewer: What do you think about the decision to drop the atomic bomb to end World War II?

Guarnere: I thank Christ for that. The generals and political bigwigs stopped sending supplies to Europe because they were getting ready for an invasion of Japan. They say there would have been more than 1 million casualties. It saved lives. Remember Pearl Harbor.

FORREST GUTH OF EASY COMPANY: 506TH PARACHUTE INFANTRY, 101st AIRBORNE DIVISION

In *Band of Brothers*, historian Stephen Ambrose states that tough Capt. Ronald Speirs had a high opinion of Forrest Guth, one of his Easy Company soldiers. Not only did Speirs think Guth was a great combat soldier, but he also appreciated the fact that the rifleman could take a file to an M-1 carbine to make it fully automatic.

Guth joined the paratroopers in 1942 with boyhood friends Carl Fenstermaker and Rod Strohl. Like Bill Guarnere, another Easy Company veteran and Pennsylvania native, Guth was an original member of the company who trained in Camp Toccoa, Ga. He made combat jumps into Normandy, Holland and survived the brutal fighting at Bastogne where he was wounded and subsequently hospitalized in England.

After the war, he earned both BA and MA degrees, and then taught school for 30 years. He and his wife, Harriet, lived in Hockessin, Del., raised two children and enjoyed their two granddaughters (one, an Annapolis graduate and a naval aviator). His hobbies included woodworking and restoring old cars.

We interviewed Guth, 85, before his death in 2009, to reminisce about Easy Company, the 101st, and his other wartime experiences.

Band of Brothers' Forrest Guth in captured German helmet during invasion of Normandy.

Interviewer: When did you decide to join the U.S. Army? Why the paratroopers?

Guth: I worked Bethlehem Steel as a machine operator and lived in a small town near Allentown, Pa. When Pearl Harbor was attacked, a couple of my crazy friends and I saw an ad in the newspaper for paratroops. The paratroops paid an extra $50 per month—this was at the end of the Depression and that was good money. We enlisted and our serial numbers were one apart from each other. In October 1942, we were sent to Fort Indiantown Gap, Pa. After that, were sent by train to Camp Toccoa, Ga. Both of my friends survived the war. Carl (Fenstermaker) died about 20 years ago and I see Rod (Strohl) from time-to-time at reunions.

Interviewer: How was Camp Toccoa?

Guth: When we arrived, Toccoa was a bunch of army tents. It was the rainy season and water ran through our tents. They ultimately built tarpaper barracks. E Company was formed and there was no other regiments training there except the 506th (PIR). Toccoa had a lot of bugs, frogs, snakes, lizards. Floyd Talbert wasn't afraid of anything that crawled and would put them in his belt. Walter Gordon was the ringleader of these shenanigans."

Interviewer: The HBO series *Band of Brothers* portrayed Maj. Richard Winters as brave, Captain Herbert Sobel as weak and Lt. Speirs as tough and crazy. What are your opinions of each?

Guth: Lt. Winters was our platoon leader early on in Toccoa. He was a good man, very athletic and in very good shape. He was a straight shooter. No drinking, no smoking, no bad women. He wasn't one of the boys—and as a good leader—you can't be. He was very reserved. He was from Hershey, Pa., and would travel with us when we had passes to go home.

Sobel was a good trainer, he insisted on discipline and doing it right. He didn't pull punches and was not popular among some of the men. He got us in great physical condition, made us follow orders and should get a lot of credit for our early training.

Lt. Speirs was a good man, good leader and honest. He wouldn't make us do anything he wouldn't do…and he did more than we did. He was a fearless leader.

Interviewer: How did the training in Georgia and in England prepare you for the rigors of combat?

Guth: The training was the best it could be. In combat, there are long hours with little food, little sleep and not too good conditions. There was not much food, so you learned to scrounge. If you didn't have it, you got it or made it. We even had guys who were butchers in civilian life. The intelligence about Normandy wasn't the greatest. The hedgerows were tall and a great place for Germans to hide.

Interviewer: What was your opinion of the German soldier?

Guth: We were trained to hate the Germans. We did hate the Germans. We had no problem shooting them. Even though I am of Pennsylvania-German descent, those feelings still carry over to today. I don't like seeing reenactors in German uniforms. After the war, companies wanted to go over to rebuild Germany, I felt uneasy about that—they were our enemy. So was Japan.

The German soldier was very good. We faced German paratroopers who were well-equipped, trained, and tough. We respected their ability and felt their equipment was better than ours. We could not wait to get our hands on some of their equipment. Their pistols were nice. I brought home a P-38 and a smaller Walther. I had a Luger, but traded it a long time ago. We noticed the Germans had burp guns and we wanted to make our carbines fire like them. It wasn't practical, as the magazines were too small. Winters had one and wanted to give me it after the war, but couldn't find it.

The Germans were just as confused as we were when we dropped in Normandy. We encountered a group of Germans having breakfast, so we shot them up and some took off. We ate their food. You ate whatever you found—better than K rations."

Interviewer: What was your favorite weapon during the war?

Guth: The M-1 was a great rifle. The carbine was not as good and didn't have good range. We also carried 3-inch mortars. We had one mortar squad in each platoon. Guys had to carry all of the pieces—the base plate, tube, and tripod. The .30 caliber machine gun also was a good weapon."

Interviewer: What was the toughest battle you were in?

Guth: Bastogne. Not only was the fighting bad, but so was the cold. We were short on heavy clothes, equipment, food. We had no air supply because of the weather for weeks. We had three men in a foxhole, only time you got out of it was to eat or go on patrol. We didn't have blankets, we had pine boughs and sometimes cardboard to keep warm. A lot of guys had frostbite. We kept warm by rotating the middle spot when a guy came back from patrol. The Germans shelled every night and the airbursts came through the trees and killed or wounded many soldiers. We lost men like Francis Millet, who was killed. Joe Toye and Bill Guarnere severely wounded.

(Don) Hoobler was in another platoon a couple of foxholes away when he shot himself (with the Luger he found). He bled to death before anyone could get to him. In today's wars, you have hospitals in the field and they can fly you out. We patrolled constantly, but so did the Germans…and we would meet on these patrols. When we attacked, the Germans were well-hidden in houses—it was not a lot of fun. We stayed out there the better part of seven weeks.

Interviewer: You went on furlough to come home after a battle?

Guth: After we were in Haguenau (France), the original members of the company, only 25 of us were left out of 138 or so, were told to put our names in a helmet. We had made it through Normandy, Holland and Bastogne. The winner got a 45-day furlough back to the States. My name was drawn—that's the reason I wasn't at Berchtesgaden. I ended up training paratroopers at Fort Benning for a Japan invasion. I never rejoined the company and never would have taken the furlough if I had known that would happen. I wrote Speirs asking him how everyone was. He wrote back and said that nothing was going on except that guys were getting drunk and killing each other by that stage in the war.

Interviewer: Is there anything in the TV series *Band of Brothers* that could have been explained more accurately?

Guth: The book was accurate. I think the series dwelled on Brecourt Manor too much. There were bridges taken and roads secured by the 101st in Normandy. On the second day of the invasion, we started to see troops coming up from Utah Beach. We were glad to see those tanks coming up. I am not much of a war movie guy. I think all of these movies make war exciting, even when it isn't. John Wayne in *The Longest Day* is an example. Movies make it easy—there are heroes and bad guys. Winters is the hero, Sobel is the bad guy. In the TV series, it showed Don Malarkey as leading the attack across the river for a prisoner in Haguenau. After he saw it, Dick Winters called me and said, 'weren't you on that mission?'

HBO treated us well, but I was not a fan of the language in the series. There was all of this yelling and cussing. We didn't use some of the words. My dad would have killed me if I had used that language. You knew your job and didn't have to yell and cuss all of the time. None of those crazy hand signals they had in the series.

Interviewer: Who do you see at reunions and what do you talk about?

Guth: We don't talk about the bad times. We talk about the good stories—trouble we got into, leaves we went on. On leave, some of these guys would get into it when they saw folks with paratroop boots on who weren't in our unit. Our jump boots were sacred.

In terms of reunions, there are a lot of good guys to remember. A great guy was Walter Gordon. He was a morale builder always playing some sort of trick. Shifty Powers was a great guy—so was Amos Taylor. Don Malarkey was a fun, but serious guy. So were Buck Compton, James Alley, Earl McClung, and Buck Taylor.

I see Eugene Perigini, who was wounded in Normandy, now lives up in Oswego, New York, at reunions. I was in Europe four times last year (for tours and museum dedications). I rode in a tank for the first time in Normandy since the war—it was a rough ride. My wife has gone with me twice on these trips. Chrysler is sponsoring a trip to a hospital in Germany this year for wounded soldiers from Iraq and Afghanistan. Seven Easy Company men are going.

Interviewer: Who was a hero to you in Easy Company?

Guth: Carwood Lipton. He was intelligent. He wasn't afraid to lead. You had a lot of empathy for those who were injured. He was wounded and stayed. He received a battlefield commission. He was in good shape, but is now passed away. He had emphysema.

Interviewer: Was it tough adjusting to civilian life after the war?

Guth: As soon as I was discharged in October 1945, I went to school. My high school principal got me into Millersville University. I later went to NYU, got another degree, and started teaching. Vets didn't want to rehash stuff...we tried to put the war behind us. We knew there were thousands of veterans who did the same thing we did. It wasn't hard to adjust. I went to Norfolk to work and back to Delaware. Now, it is time to think back about a few things. I had good friends who came back crippled or didn't come back. More than 50 from Easy Company didn't come back.

Interviewer: How do you want people to remember Easy Company?

Guth: They were a bunch of great guys. We lived together for more than three years and we knew we were the best. Along with my wife, they are the best friends I ever had...and still are my best friends.

Interviewer: Is there anything about society in that era that you think is gone—and think should be restored? Any life lessons you learned during the war?

Guth: Discipline. Parents should be more involved in disciplining their children. We didn't have all of the gadgets these kids do today. With all of the television and other goodies kids have today, there is no wonder no one wants to enlist in the volunteer army.

Growing up during the Depression, my family was not wealthy. My father was a railroad worker who also had jobs at a cement plant and on a road crew. Just like in the military, we had to make do, we had to improvise.

MYRON BERRICK: SILVER STAR RECIPIENT SERVED IN BOTH THE PACIFIC AND EUROPEAN THEATRES

Soon to become one of the elite American armored units of World War II, the 10th Armored Division (nicknamed the "Tigers") was activated on July 15, 1942 at Fort Benning, Ga. It continued its training cycle in the U.S. until being shipped to France in September 1944, whereupon it became a key combat unit in Lt. Gen. George S. Patton's Third Army.

American WWII armored divisions were complicated organizations composed of three combat commands, A, B, and R. Each contained tank, armored infantry, field artillery and tank destroyer battalions, supplemented with divisional mechanized cavalry and engineer units. The 10th was no exception. The Tigers remained with the Third Army during its entire combat existence, although it was often moved among Corps as the tactical situation demanded.

The 10th AD entered combat near Mars-le-Tour, France, and helped take the fortress city of Metz in bitter fighting. It later was used to force crossings of the Moselle and Saar rivers. During the Ardennes counteroffensive, the division rushed north to relieve members of the 101st Airborne at Bastogne, although the unit does not get much credit for its participation in the Battle of the Bulge. The Tigers then fought to capture the ancient city of Trier and numerous other German cities. It ended its war in Imst, Austria, near Innsbruck, where it settled in as an occupying force.

Myron Berrick during World War II

We interviewed Myron Berrick, then 85, a 10th Armored Division officer, who served in both theatres of World War II. He was awarded both the Silver and Bronze Stars, as well as the Combat Infantryman Badge. Berrick, who lived in New York, was married to Frances for 63 years; they have a son, Lawrence, and a daughter, Charlotte. He passed away in 2014.

Interviewer: Why did you decide to join the Army?

Berrick: I didn't have a choice. I was drafted. I was going to NYU on a fencing scholarship. When the coach left, I became the freshman fencing coach. When I was drafted, I interviewed with a sergeant who asked me what I did in college and I told him I taught fencing. He went over to a big book to look at, then came back to tell me I was going to be assigned to Camp Lee, Va., the Carpenter's School. I later told the sergeant major in charge of the school what I did in college and he nearly fell off his chair. He asked me to teach him and his wife how to fence.

Interviewer: Where you reassigned?

Berrick: I was sent to the Quartermaster Corps and passed the Officer Candidate examination at Camp Lee. I went to a school in Nebraska, the Motor School, then went to San Francisco as a DUWK (2 1/2 ton amphibious truck) officer. Although San Francisco didn't have a beach, we all conducted harbor boat patrols. I did do a tour as a security officer on a Liberty Ship for a short time in the Pacific in 1943.

Interviewer: Did you stay in the Pacific Theatre?

Berrick: No, I was transferred to a truck company in the Texas National Guard at Fort Hood. It was a culture shock. I was from the North and was a Jew. They would say I was still a white man, even if I was 'only a Jew.' I also was put in charge of training Nisei troops, who did a hell of a job. On Saturday nights, the Japanese kids just sat around—they couldn't be served a beer in town. I put them in a truck and told a bartender to serve them, but he refused. Then I told him I wanted 18 beers and he said he could only serve me one.

During this time, they had a directive that would transfer any 2nd Lieutenant who wanted to join the infantry, which I did. I went to Fort Benning, Ga., for additional training. I saw another officer who had been a coach at NYU who requested me for a special unit that worked behind enemy lines. I wanted no part of that, but it turned out that later I would be behind enemy lines anyway. I was transferred to the 10th Armored Division because they needed a motor officer in the 54th Armored Infantry.

Interviewer: When did you arrive in theatre?

Berrick: We arrived in Scotland. We went down to South Hampton and landed at Le Havre, France, in January 1945."

Interviewer: What was your opinion of Gen. Patton?

Berrick: The 10th Armored was with Patton for a while, then elements of the 17th Panzer Army ambushed one of our companies. When we finally got to our soldiers, we found bodies that were shot and bayoneted, like Malmedy (*editor's note: the Malmedy Massacre in Belgium occurred during the Battle of the Bulge when 84 U.S. prisoners of war were machine gunned*). Ironically, their equipment was in neat piles. Patton was a sentimental guy. He wanted us to be attached to a unit that would find [the Germans responsible]. We did and wiped them out. I still have one of their battle flags.

Patton was a mean son-of-a-bitch. He was a brilliant general, there's no doubt about that. He picked our unit to fight a battle near Landau, Germany. It was a tough time as 17 German 88s were pointed straight down at us. He insisted that we take the town. We took the town after a bitter fight, but I will say this about the bastard, he was right at the front.

After the war, I was at a dance in Garmisch-Partenkirchen for majors and above, although I was only a first lieutenant. He walked up to a major dancing with a nurse and pushed the guy out of the way and started dancing with her.

Interviewer: Do you have one memory that stands out about combat?

Berrick: Only how terrible it was. In one battle, we were about to attack a town in Germany, we were waiting for a time-on-target artillery barrage where every gun in the division fires at different times, but the shells land at the same time. It was in response to President Roosevelt's death. In the town, there were an outer ring of regular Wehrmacht troops and civilian defenders. The second ring had SS troops. We went in and annihilated the town.

There are things that are hard to talk about. We were at one of the smaller concentration camps and we noticed chains that placed prisoners in such a way that allowed trained attack dogs to jump up and bite them in certain areas.

Interviewer: Did you stay to be part of the occupation forces?

Berrick: I stayed in theatre for a year after the war ended. I was a Provost Marshal in the port city of Bremerhaven. They also put me in charge of guarding the division headquarters. On Hitler's birthday, we heard that a Nazi group, called the Werewolves, were celebrating and we were ordered to stop them. We walked across an open field and found kids--ages 10 to 14. They did have a panzerfaust (anti-tank weapon). We put them in trucks and brought them to the colonel in charge. He scared and berated them. We also noticed that families would go up into the woods with picnic baskets in the Redoubt (where it was alleged that the Nazis had a stronghold) area. An hour or two later, a mother and father would return with a young man in civilian clothes. We realized that the young men were SS hiding out in the woods. We got a lot of prisoners that way.

I finished the war as a first lieutenant. There was a rule than any second lieutenant who served 18 months or more was automatically promoted to first lieutenant.

Interviewer: What did you do after the war?

Berrick: I worked at various jobs and then got into clinical psychology. I used the GI Bill to finish my college degree at NYU in mental hygiene. I worked at the VA Clinical Psychology program for seven years. I started my own mental hygiene office and worked at Brooklyn College for 37 years.

KEN BLACKSHAW: WORLD WAR II B-17 PILOT WHO SURVIVED 35 MISSIONS OVER GERMANY

Ken Blackshaw flew 35 combat missions over Germany during World War II and rubbed elbows with some of the biggest names in Air Force history—the stuff that movies are made of. In fact, the movie and television series *12 O'Clock High* were based on Blackshaw's unit, the Eighth Air Force's 306th Bomb Group.

Blackshaw served in the Air Forcefor 21 years, retiring in 1963 as a lieutenant colonel. While on duty in the Pentagon, he maintained his pilot proficiency by flying out of Bolling Air Force Base. Following his six Pentagon years, Blackshaw was assigned in January 1952 to the Third Air Force in London as Special Assistant to the Commander for Protocol. He maintained his pilot skills during this period with assigned flights throughout Europe and Africa.

It was during this time that U.S. Strategic Air Command was rotating bombers to bases in England. The 10-engine (6 props and 4 jets) B-36 was a first-line operational bomber in 1953, Blackshaw remembered.

The Convair B-36 Peacemaker was the largest mass-produced, piston-engine aircraft ever built. It had its design origin in 1941 when the Army Air Corps saw the need for a long- range bomber that could reach targets in Europe from North American bases. With the survival of Britain and the occupation of southern Italy, the USAAF decided that the strategic bombardment of the Nazi Reich could be better undertaken with the smaller, shorter range B-17s and B-24s. As a result, the first B-36 didn't fly until 20 August 1946. It was so large (requiring a flight-crew of 15) that initially only three Air Force bases in the U.S. could accommodate it.

The Peacemaker remained in service with the Strategic Air Command until replaced by the B-47 Stratojet in 1955. Only four B-36s (non-operational) of the 384 built are still in existence.

England's Prince Philip had expressed an interest in seeing a B-36. A press tour was arranged that included the prince. Blackshaw was the pilot of the press plane for the mission. While the prince was in the cockpit of the B-36, Blackshaw decided he would take a look at the tail section, which was separated from the cockpit by long tube that one could use to slide from one end to the other. After viewing the tail section, Blackshaw went to step down the ladder to get back to the ground. There was an object in the way: Prince Philip's head.

Blackshaw said that prince looked up and said, "Hello, I'm Philip" with a strong British accent. Blackshaw, who quickly got out of the way, proceeded to bang his head on the aircraft while the prince entered.

Blackshaw's other assignments included Strategic Air Command headquarters in Omaha, Neb.; commander of Chicksands Priory Air Base north of London; U.S. Air Force Europe Headquarters, Wiesbaden, Germany; 301st Air Refueling Squadron, Barksdale AFB,

La.; and Goose Bay Air Base, Labrador. He served with or knew famous Air Force generals Archie J. Old, Jr., Curtis LeMay and Carl "Tooey" Spaatz.

After military service, he worked for 18 years as the Director and Public Affairs Officer for the Combined Federal Campaign for the Secretary of Defense, ending his tenure in 1997. He also served as the National Press Club's American Legion Post 20 commander in 1994 and 1995.

Blackshaw, who passed away at 94 in 2012, resided in McLean, Va. Married more than 35 years to Elizabeth McNally Blackshaw, he is survived by his son, three grandchildren, seven great-grandchildren, and two great-great grandsons.

Interviewer: When did you join the U.S. Army Air Forces?

Blackshaw: I was working for Nantucket *Inquirer and Mirror* in 1942, when I went to the Navy Yard in Boston. I passed my physical, but the doctor asked me if I had any allergies. I said no, just hay fever. The Army Air Corps rejected me on the spot. I went back two months later and passed everything again, but this time, when asked about allergies, I said, 'no, sir.' They swore me in 'on the spot.

Interviewer: Where did you do your training?

Blackshaw: In January of 1943, they called me to active duty. I reported to Atlantic City, N.J. While the Army wasn't ready for us, we did eat in a chow hall to the sound of a live orchestra headed by Capt. Glen Miller. I got to meet him later when he got to England. He wasn't very talkative, but played a decent game of snooker (billiards).

They didn't send me to pilot training right away. I was sent to attend college at the University of Pittsburgh. I spent four months there living, learning and sleeping in the 39-story main university building. I was then sent to a Classification Center in Tennessee, where I was selected to be a pilot, in lieu of navigator and bombardier.

In Orangeburg, S.C., I learned to fly solo in a Stearman biplane. I used to get air sick, which was a good way to get kicked out of the cadet program. To get around this, I learned to fly upside down and let it go whenever I had to upchuck. Only a leg strap was holding me in the airplane.

Interviewer: Did you go overseas after pilot training?

Blackshaw: At 26, they told me I was too old (for combat), so they made me an instructor. One day, in the orderly room, I saw a flyer for a requisition for B-17 pilot transition. I scribbled in 'Blackshaw' and the next thing I knew I had applied for it and was selected.

Interviewer: Where was B-17 training?

Blackshaw: They sent me to Columbus, Ohio, for pilot transition training. After that, I went to Rapid City, S.D., to assemble a crew and go through 10 weeks of difficult training. It was tough as I made them cope with every imaginable scenario for combat. The crew was sent to Lincoln, Neb., to pick up our brand new B-17 to fly to England. One of our crew members was from western Iowa, not too far from Lincoln. We flew over his farm and dropped flares and smoke bombs. The folks on the ground waved back. When we prepared to go overseas, a superior officer asked me if I, Lt. Blackshaw, had dropped the flares. I said yes. Apparently, I was the only pilot who had tested this type of equipment before. He said that I could not go overseas unless I paid back the government $18.73 for the flares I dropped. I had to pay the bill to go overseas.

Somehow, we got 'lost' flying to New Hampshire and found ourselves over Nantucket, population 5,000 to 7,000 at the time. We gave the town the greatest fly over! My wife and son waved at me, I was told later. We went on to Goose Bay, Labrador, spending a few days there getting airplane parts. Then we went to Iceland, where we spent one night in a tent, with 34 blankets between us.

Interviewer: What happened when your B-17 landed in England?

Blackshaw: We landed and had to take a train from London to Bedford, England which was the closest to our base, Thurleigh, home of the 306th Bomb Group, 423rd Squadron. We were told to bring bourbon—and we had several bottles with us. Scotch was scarce.

Interviewer: What was your first mission like?

Blackshaw: We flew our first mission in October 1944. For the first two missions, I was required to fly as co-pilot before taking my own crew on a mission. On the first one, I noticed all the little squares on the B-17 and asked what they were for. I was informed that they covered up the holes made by flak from previous missions. When the mission was over, they counted 76 new holes. That's where we had been hit.

Interviewer: The 306th had a distinguished career. Where did you bomb?

Blackshaw: We bombed all over Germany. We bombed Berlin twice. Hitler swore we would never bomb Berlin. We hit Würzburg three times. The flack was so thick that you could almost walk on it. Berlin's flack was not that heavy, but very accurate. We bombed the Nazi scientific center at Peenemünde. After the war, I told the head of the center that we had bombed them. He had no hard feelings, and said, 'you did a good job.' I was lunching with him at the New York World's Fair in 1964.

We flew as the No. 2 ship during many missions, had the Norden bombsight and had radar. On many raids, we would get up at 3 a.m. for the lead crew briefing, and then to the crew briefing at 4 a.m., to prepare to take off at 7 a.m.

Ken Blackshaw's B-17, "Gray Lady."

Interviewer: Any missions stand out as harrowing?

Blackshaw: During the Battle of Bulge, among many others, we flew on three missions, 1-2-3 January 1945. We took off in the dark and landed in the dark. Nearly 2,000 airplanes also did the same thing. There was a very short daylight span.

Sometimes we were sitting game. I recall flying over Holland, where the Germans isolated anti-aircraft batteries. A bomber on our flank was hit by flak in its bomb bay. The plane completely disintegrated.

Interviewer: How were the relationships among the crew?

Blackshaw: The crew was very dedicated. We had a crew member from Utah, two Catholics, one Jewish airman and Protestants. Our bombardier was a freshman at Notre Dame and married. Our radio operator, Ed Tutten, should have gotten the Silver Star as he saved the life of one of our gunners who was shot and bleeding furiously during one mission. He administered first aid with a tourniquet. The member he saved was Chuck Yeager, not *that* Chuck Yeager, though I met the famous one later. My Chuck Yeager just passed away recently.

We also loved the fighter pilots--they protected us. Occasionally, we would lose an engine and when we shot a red flare, would very quickly see a fighter pilot flying close to protect us.

Interviewer: Do you still keep in touch with the crew?

Blackshaw: I didn't right away. Out of the 10 crew members, eight are now deceased. I kept in close contact with them over the years. We were the 306th Bomb Group, first to bomb Germany, first to earn the Medal of Honor. We flew many of our missions together.

Interviewer: What are your thoughts about Gen. Curtis LeMay?

Blackshaw: I knew him personally. In retrospect, I am a great admirer. The people of Omaha loved him. He used to fly into London when I was there. My general was a race car enthusiast who would go to France for big races like Le Mans. They both bought English basic cars with no engines. LeMay was a good friend of the general and would occasionally attend the races. One time, we were told that LeMay had a top, top secret guest who turned out to be Arthur Godfrey.

Interviewer: Did you end up staying in Europe after the war?

Blackshaw: We flew a lot in all three corridors into Berlin. In general, Russian MiG pilots would closely monitor us. Flying conditions were poor. A lot of fog and zero visibility.

After being a member of an elite (General Tunner's Pink Card, the only ones allowed to carry passengers) crew for four years in Europe, I was reassigned to a unit of Strategic Air Command. I was sent to Barksdale, La., where I was an aircraft commander of a crew. They did make me the administrative ffficer of the 301st Air Refueling Squadron, which allowed me to wear a 35 milimeter gun when we reflexed to Goose Bay, Labrador, which was often.

The 7 a.m. roll call every morning didn't suit me much. I did refuel Jimmy Stewart's B-47 during his reserve training. Later, my wife Betty and I met him and his wife Gloria in Hollywood at an event. They were very nice and we had a great conversation with them.

Editor's Note: Blackshaw was a member of these organizations: The National Press Club (Silver Owl); American Legion, Post 20, NPC; Knights of Columbus (PGK); Touchdown Club; Military Officers Association; Order of the Daedalians; Disabled American Veterans; Tuesday Musical; Aviation/Space Writers Association; National Aviation Club; Eighth Air Force Historical Society, local and national; and the Aero Club of Washington.

RICHARD BAYLES: A LIBERATOR OF DACHAU

The liberation of Dachau concentration camp by U.S. forces in April 1945 exposed the brutality of Nazi Germany to the world. One of the liberating divisions, the 42nd Infantry "Rainbow" Division, a National Guard division created from various state guard units, had an illustrious combat record in World War I. At its July 1943 federal activation ceremony, the new division commander, Brig. Gen. Harry Collins, echoed its former commander Gen. Douglas MacArthur's earlier sentiments on the 42nd Infantry Division's unique status when he said, "The Rainbow represents the people of our country."

The 42nd landed in France in December 1944 assigned to the 7th Army. It advanced through France and entered Germany in March 1945. It was during the 42nd Division's rapid advance through Germany in April 1945 that they, along with the 45th Infantry Division, liberated the infamous Dachau concentration camp.

While not in combat as long as other divisions, the 42nd Division still established a solid record while taking 3,971 casualties. It was first in VI Corps to enter Germany, first to penetrate the Seigfried Line and first into Munich. Rainbow Division soldiers had seized more than

6,000 square miles of Nazi-held territory during their march across Europe. The Division ended the war serving as occupation forces in Austria. It was inactivated in June 1946.

We interviewed Bayles before his death in 2011. Dick, 90 at the time of our interview, was a lifetime Denver resident, having attended Denver's South High School with friend Keith Ware, later a major general and World War II Medal of Honor recipient who was killed in Vietnam. Dick worked for Montgomery Ward from 1933 to 1980. He lived in Denver with his wife, Rose. He had a son, Richard, Jr., who is a physician, and two grandchildren.

Interviewer: How did you enter the Army and where? What was your job?

Bayles: I was drafted and entered active duty from Denver. I was first sergeant of Headquarters Company, 42nd Division Artillery, 42nd Infantry Division.

Interviewer: Where did your unit invade France?

Bayles: We landed near Marseilles. We then advanced through the Rhone Valley.

Interviewer: The 42nd was in France's Vosges Mountains. Can you tell us about the campaign?

Bayles: It was very rugged. It was a heavily forested area. We would advance two miles, then retreat four miles.

Interviewer: Your division crossed the Rhine River and poured into Germany. What were your objectives?

Bayles: We crossed the Rhine north of Strasbourg, France, attacked the Siegfried Line, then the unit went into Würzburg, Germany, from April 2-6, 1945. The Nazis had a sign that said, 'Heil Hitler,' on the castle overlooking the city. Our Rainbow Division engineers crossed it out and painted our division flag over it. There were big rainbows painted everywhere. From Würzburg, we took Schweinfurt on April 9-11. After that, the division moved to Furth on April 17-19 and then Nuremberg.

Infamous Dachau Concentration Camp near Munich, Germany.

Interviewer: When did you receive orders to go to Dachau?

Bayles: We were in the Nuremberg Valley—and Munich was about 81 miles away. We knew we were going to go to Munich, where Hitler started the Nazi movement. Dachau was 15 kilometers away from Munich. We liberated Dachau on April 29 and 30. We also were in Linz, Austria, where the Germans had gas chambers and crematoriums.

Interviewer: What were your feelings as you entered Dachau for the first time?

Bayles: The first thing you see when you enter Dachau is the train full of bodies. The Germans said that they would execute their prisoners, rather than to have them liberated. I can tell you I felt mad. Angry at what they did to people.

Interviewer: Were there any German guards left?

Bayles: There were quite a few German SS troops left. We held them to do hard labor.

Interviewer: Can you describe Dachau?

Bayles: The layout of the camp at Dachau was similar to other Nazi camps at that time. There were long rows of barracks. Now there are monuments from several religions and chapels including Jewish, Catholic, Protestant and other faiths.

Interviewer: Do you have a stand-out memory during this time?

Bayles: I was in the back of a 2 ½-ton truck and was fined by Gen. Patton for not wearing a necktie in Munich.

Interviewer: Where were you and what were your duties after the surrender of Germany?

Bayles: We had occupation duty in Salzburg, Austria, where we rested and refitted. We thought we were going to go to Japan because the war in the Pacific was still under way. We had only been in combat for five months. The other infantry divisions with us during the invasion of southern France were the 45th and 3rd divisions. They had been in combat for 500 days in North Africa, Sicily, Italy, France, and

Germany, so we thought we would be the next to go [to Japan]. I was released from active duty on March 16, 1946.

Interviewer: Would you recommend a book for our readers?

Bayles: *Legends of Dachau.* You can find it at your local library.

CHARLES NORMAN SHAY: HERO OF D-DAY

Unlike many divisions landing on D-Day on June 6, 1944, the U.S. First Infantry Division, nicknamed The Big Red One, was a battle-tested formation that had seen action in North Africa and Sicily.

The Big Red One landed in North Africa in November 1942 as part of Operation Torch. In July 1943, it spearheaded the invasion of Sicily, codenamed Operation Husky, where it faced tough German and Italian resistance.

During D-Day, the division landed in the first waves at Omaha Beach, suffering heavy casualties. It secured the beachhead and then advanced inland. It fought across France, Belgium and Germany, seeing heavy fighting in the Hürtgen Forest and Battle of the Bulge.

After the German surrender in May 1945, the division remained in Germany as part of the occupation forces. The elite division earned 17 Medals of Honor, nine posthumously, during World War II.

Charles Norman Shay, born June 27, 1924, was a member of the famous First Infantry Division, in the Medical Detachment. He was attached as a platoon medic to F Company, 2nd Battalion, 16th Infantry Regiment, when he landed in the first wave at Omaha Beach.

In an interview with Mike Van Buren for the Library of Congress, Shay, a Native American and member of the Penobscot tribe, grew up in Indian Island, Maine. After he entered the U.S. Army, he was sent to Camp Pickett, Virginia, for basic training, and then to Fort Benjamin Harrison, Indiana, for follow-on medical instruction. He was sent to England, onboard the Queen Elizabeth (which was turned into a troop ship), to join the First Infantry Division.

On June 6, 1944, he assisted his wounded comrades on Omaha Beach with bandages, splints and administered morphine. He saved the lives of several soldiers, trapped in the rising tide, from drowning. For this and other acts of heroism, he was awarded the Silver Star.

Sadly, one soldier he could not save was his friend, Pvt. Edward Morozewicz, who died on the beach. In an Associated Press story from 2024, Shay said he tried to comfort his friend, but knew he was dying from a stomach wound he could not treat.

Shay fought through the Normandy Campaign, during the battles at Aachen, the Hürtgen Forest and Battle of the Bulge. He told Van Buren that he did not try to be close to anyone in his unit as they were often killed. He did look up to a fellow Penobscot tribal member, Melvin Neptune, who had served in many of the division's battles.

In March 1945, Shay was part of a patrol near Auel, Germany, that was surrounded by 20 German soldiers—and forced to surrender. He marched miles by night to Stalag VI-G. Ultimately, he was liberated on April 12, 1945, by American troops.

According to the Van Buren interview, Shay's family never knew he had been captured, as he was listed as missing in action. They only knew of his fate when he showed up at the family's house in Boston after the war.

After the war, he had a few jobs, but missed the military and chose to reenlist in the Army, serving in combat again in Korea with the 3rd Infantry Division. He served as a combat medic in the 3rd Division's 7th Infantry Regiment.

Charles Shay with Margaret Heesen and Brigadier General (ret.) Ted Shuey.

After the Korean War, he reenlisted in the U.S. Air Force, retiring in the 1960s, and working in Austria with the International Atomic Energy Agency. He married and had a son.

A few years ago, he was awarded the Légion d'Honneur by the French government. In addition, in 2017, the Charles Shay Monument was unveiled on a dune near Omaha Beach. The monument, which features Shay's bust and a sculpture of a turtle that is sacred to the

Penobscot tribe, commemorates all Native Americans who landed on D-Day.

For this book, Margaret Heesen interviewed Shay in late 2023. He now lives in Normandy, France, and turned 100 in 2024.

Interviewer: What was your motivation for joining the military?

Shay: I had no choice. I was drafted.

Interviewer: What was your experience during training?

Shay: The military training was something I never experienced before, it was all new for me. I was sent to Fort Devens, Massachusetts. Once I was processed, I was sent to Camp Pickett, Viriginia, in the summer of 1943 for my basic training. There I received my basic introduction to the military service and trained for how to march, how to dress, how to conduct yourself and learn about your life in the military service.

Interviewer: What follow-on training occurred?

Shay: I was selected and was sent for further medical training at Fort Benjamin Harrison, northeast of Indianapolis, Indiana, where I stayed for about eight weeks of intense training. I did not choose to become a medic—the Army needed more medics so they put you in there.

I was trained in further medical tasks, got some knowledge about anatomy and became a circulating nurse in the operating room of a field hospital. I was trained to prepare the operating room and surgical equipment, cleaning and sterilizing equipment, assisting patients from and to the operating room.

Interviewer: What other duties did you have?

Shay: I performed work of a scrub nurse and circulating nurse assisting in operations, hold patients on the operating table for injections before they were operated on and I was also responsible that no item such as bandages has not been left in the body of the patient by counting them and make sure all items have been eliminated before closing the wound. I was trained to become a surgical technician. Once this training has been completed by early fall, I thought I had a good position and will work in a field hospital. Unfortunately, it never happened.

Interviewer: What was the most difficult thing you experienced during your time in combat?

Shay: "The invasion of Normandy, which was a quite large operation to defeat the enemy on the beach, probably equal to what we experienced later on in the Hürtgen Forest, but it was a smaller operation.

Interviewer: What are your feelings about being a Native American during World War II?

Shay: Nothing. I never felt different from my comrades.

Interviewer: How was your POW experience?

Shay: I was treated like other prisoners. We were in the thousands at the Stalag.

Interviewer: What was the most rewarding thing you experienced during your time in combat?

Shay: Defeat the enemy on D-Day.

Interviewer: How did military service in World War II change you?

Shay: It made me aware we have to be prepared to defend our country against the enemy who wants to change our way of life.

Interviewer: How should the war be viewed by today's younger generations?

Shay: They should be prepared to defend their country if ever it becomes necessary again.

WILLIAM GEIGER: AN AMERICAN IN THE RAF

Before the United States entered World War II, many Americans volunteered for service in both the Royal Air Force and the Royal Canadian Air Force. As a result of losses during the Battle of Britain, the RAF was critically short of pilots and eager to enlist flyers of all nationalities.

Of the thousands of applicants, 244 American pilots eventually flew for Eagle Squadrons Numbers 71, 121 and 133 of the Royal Air Force Fighter Command. Those units went on to destroy 73 German aircraft, but lost 82 of its pilots, five of them British. Eventually, the Eagle Squadrons were absorbed into the U.S. Army Air Forces.

Rob Woodard interviewed Bill Geiger, who served in Eagle Squadron, about his decision to join the RAF and his wartime experiences. Geiger, whose call sign was "Deacon 33" for the "Deacon 71" Squadron, was eventually shot down over the English Channel and held prisoner at Stalag Luft III, the Nazi POW camp later made famous in the movie, *The Great Escape* (1963).

His father, William D. Geiger, an early aviator, is buried at Arlington National Cemetery. Bill's uncle, Harold Geiger, a West Point graduate, died in a military airplane crash—Geiger Field in Spokane is named for him.

Bill Geiger resigned from the Royal Air Force after the war. He returned to California where he owned a successful real estate business. He lived in Santa Barbara before his death in 2006 at age 86.

Bill Geiger in World War II.

More information on the Eagle Squadrons can be found at the National Museum of the U.S. Air Force website: http://www.wpafb.af.mil/museum/history/prewwii/es.htm.

Interviewer: When you volunteered to join the Royal Air Force before the United States had entered the war, what was your motivation?

Geiger: I wanted to fly. I had learned to fly in college and I had a private license, but I didn't have enough college credit to get into the U.S. Army Air Corps. At the time, you had to have two years of college and I only had about a year-and-a-half.

Interviewer: How did you learn about the opportunity to join the RAF?

Geiger: There were rumors around the place where we were taking lessons, Alhambra Field in Alhambra, Calif., that the RAF was looking for people to join an all-American squadron they were calling 'Eagle Squadron.' My friend, Tommy Wallace, told me that there was a place in Hollywood where you could sign up to join Eagle Squadron, so we went up there and did it. It sounded like an adventure.

Interviewer: How old were you?

Geiger: I joined in the fall of 1940. I was 21 years old.

Interviewer: So after you signed up, where did you go? What was your training?

Geiger: We went to Glendale, Calif., to a flight school run and taught by civilians, who were ex-Army Air Corps pilots. The British government paid for our training.

Interviewer: Were you British officers at this time?

Geiger: No. We had not yet been commissioned so we wore civilian clothes and did not train with the Army Air Corps, which was on the same property as our school--although we ate the U.S. Army's food and stayed in Army barracks.

Interviewer: So when did you officially become British officers?

Geiger: When we finished our training we reported to a freighter to take us to England, as soon as we boarded that boat, we were commissioned.

Interviewer: What was your rank and pay?

Geiger: We were commissioned as "Pilot Officers" which was equivalent to a second lieutenant in the U.S. Army Air Corps. We were paid 21 pounds a month, which was about $84. In those days, that money went pretty far, a pound was worth about $4. In addition, the English people always wanted to help us, so there were many occasions when we would go into a bar to have a drink and they wouldn't let us pay for it!

Interviewer: Once you were in England, where were you posted?

Geiger: Martlesham Heath, which was a Royal Air Force base."

Interviewer: What type of aircraft did you fly?

Geiger: I started out flying 12-gun Hurricanes and then we went to the Spitfire. Although Spits only had four guns, they were cannons. If you hit anything with those cannons, that was adequate.

Interviewer: What were your flying duties?

Geiger: We were flying close escort on daylight bombing. The British decided that after they slowed Hitler down, and he had not invaded England, that they would show the French that they were now on the offensive. So, the Brits would take six to 12 Blenheims, which were ill suited for daylight bombing, and send them over France so that the French could all see them. And, these bombers would bomb German-occupied areas. Our job was to fly close escort because the bombers were incapable of defending themselves. Usually, I flew behind the bombers, protecting the rear of the plane. The German ME 109s would come up behind us and try to break through to get at our bombers. When they couldn't get through, they would break up or break down and go away, that would be the end of it.

Interviewer: What happened on the day you were shot down?

Geiger: We had been over to France and we brought the bombers back over the English Channel. We had finished the mission and we were over the Cliffs of Dover. I heard that a guy behind me was having trouble with some German 109s. So, figuring that the bombers were home free and safe, I circled back to see if I could help. I got back a little way and there were three 109s waiting for me.

Interviewer: So it was a trap of some sort?

Geiger: Well, that's the way the Germans played… they'd watch for somebody to get in trouble and then wait for someone else to come to help -- and then they'd get you. So they nailed me and I had to bail out into the English Channel. I was in the water for pretty close to five hours when a German patrol boat picked me up.

Interviewer: Whatever happened to the guy you went back to help?

Geiger: (Laughing) I don't know, I think he got shot down, too. I was too damn busy with my own thing to worry about him!

Interviewer: So they haul you aboard--then what?

Geiger: They took me to Saint-Omer, France, and the next night I went to a town called Lille. I was concerned because we had been told that if the Germans caught Americans they would shoot us as soldiers of fortune. I get to Lille and they wake me up at I guess about 5 a.m. It was cold and the streets are deserted. There are several guards and myself and they start walking me down the street. I was thinking to myself that if ever I saw a firing squad, this was it. I figured that I was not just going to let them shoot me, so I was looking for a place to run to at least give them a moving target.

Interviewer: You were the only prisoner they had at this point?

Geiger: Yes. So eventually we get to the outside of town and there was a lorry. And there were about another six soldiers there. If I ran, I would have to run across a field while being shot at by now about 10 guys. They told me to get in the truck. (Laughing) I got in the truck.

Across from me sat a German soldier. He spoke to me with a Brooklyn accent and asked if I were American. When I told him I was, he assured me that I was going to be okay. A wave of relief came over me as I realized just how nervous I had been that they were going to shoot me. I said: "You don't speak with a German accent, you sound like you're from Brooklyn!" He told me that he and his family had moved to America when he was a baby. Hitler had offered these former Germans practically a free ticket to come back to see how marvelous things were under the Third Reich. When they came back, he was drafted into the German army. He said he was resigned to the fact that he was not going back to the States anytime soon. There was nothing he could do about his situation. When I asked if there was anything I could do about my situation, he told me that "you had better go where they're sending you" as he looked from side-to-side to see if anyone had been listening. Ultimately, I wound up at Stalag Luft III.

Interviewer: What was life like in Stalag Luft III?

Geiger: Well, we spent a lot of time playing softball and walking around what we called the "circuit." The food was terrible, but we got Red Cross packages. To this day, I still consider spam to be as good as a sirloin steak. That was a real highlight--to get a slice of spam.

Interviewer: How were you ultimately liberated?

Geiger: In February of 1945, it started to look like the Russians were getting close to freeing us. So the Germans marched us out of there. (*Editor's Note: Geiger was moved from Stalag Luft III just before the "Great Escape" occurred; although he had participated in the digging the escape tunnel, he was not there for the breakout*). I ended up in Nuremburg, Germany. A lot of guys got sent to Moosberg [Stalag VIIA]. When it looked like the Allies were getting close to Nuremberg, they marched us out to Moosberg. I don't think I was in Moosberg much more than a week and then Patton's 3rd Armored Division liberated us."

Interviewer: What was that experience like?

Geiger: The Army drove a tank right through the front gate. There was a book written about that, *The Day I Saw Ten Thousand Men Cry*.

It took a while for us to get processed out. I was put in charge of a warehouse in Moosberg that held a lot of Red Cross food. I think I owned half of Moosberg by the time I was done. When it came time for me to leave, I said to Captain Stocker, Patton's supply officer, "I tell you what, Jim, I know you're short-handed, but I'd be glad to hang around for a couple more weeks." He declined and told me to go home."

BENJAMIN BERRY: PENNSYLVANIA SOLDIER FOUGHT GERMANS, DISCRIMINATION

Benjamin Berry, now 100 and living in Germantown, Pennsylvania, was a gung-ho teenager when he entered the U.S. Army during World War II, serving from 1943 to 1945.

According to a 2024 article in the *Chestnut Hill Local*, written by Kristin Holmes, Berry grew up in the Crestmont section of Willow Grove. He graduated from Abington High School in June of 1943. After entering the Army in July 1943, he was in basic training at Fort Frances E. Warren in Cheyenne, Wyoming.

The U.S. military was segregated during World War II, with black soldiers serving in units usually commanded by white officers. In an interview with Margaret Heesen in 2023, Berry said he was assigned to a segregated quartermaster unit that transported supplies and operated facilities for soldiers to shower when they came out of battle. Berry, a member of the 863rd Quartermaster Fumigation and Bath Company, also drove a truck that transported supplies.

He soon learned the inequity of being a black soldier during the war as the Army gave him a rifle, but didn't provide ammunition. That changed when his unit, which followed the infantry, was sent to the Battle of the Bulge. He and other black soldiers were finally given ammunition to help fight off the Nazi counterattack in the Ardennes Forest.

Then and now, World War II veteran Benjamin Berry.

Berry, who left the service with the rank of corporal, landed in France after D-Day. He served in the Normandy Campaign, Huertgen Forest and the Battle of the Bulge, among other battles. He was with the unit when it helped to liberate Pilsen in the Czech Republic. After the war, Berry went to school on the GI Bill. He married Lois Miller Graves and raised four children in Pennsylvania.

Berry recently received a citation and medal from the Grand Duke of Luxembourg to commemorate his war service. He also was named to the U.S. Army Quartermaster Hall of Fame. In addition, in March

2024, he was honored with the Légion d'Honneur for his service in Europe during World War II.

Interviewer: Why did you join the U.S. Army?

Berry: I had no choice, I was drafted. Uncle Sam said, 'I need you,' so you went. In fact, in some cases, the military came after you and put you in jail if you did not go when called. Only reason you did not go is if you had a medical excuse. We did not try to make up a fake excuse not to go because you'd be considered an F4, a reject, a wimp by friends/peers. We didn't try to pull any strings either to get us off the hook so we didn't have to go. The army designated the men in different classes: Classes A, B, C and D. If you were considered one of these, then you're able to go.

Uncle Sam, and even the girls in the community, did not want anyone who was a Class F4. Other soldiers, those who were accepted, called you a 'Jodie' if you weren't physical able to go into the war. Even having flat feet did not keep you out of WWII because at least you could peel potatoes.

Interviewer: Did anyone else in your family serve during the war?

Berry: My brother 'June' (James) was drafted into the Army, but I don't recall which area. I think the Pacific somewhere. But I ran into him one day at a recreational area in France. There were many other military outfits there too and I just happened to see June. I sure was excited and happy to see him. 'Hey what are you doing here?' was our first comment to each other. We hung out an hour or so, which was great. Don't know anything about him, or what unit he served—or even what we did while we were there socializing.

Interviewer: What did you learn during training?

Berry: To follow orders and not to be last. That includes for reveille call in the morning before breakfast. During the war, one of my jobs was to escort soldiers to the infirmary. Make sure we did what our job description was to a 'T,' even finding water in a stream and get it for the troops if need be. No ifs, ands, or buts. No excuses!

Interviewer: How did the Army making basic training tough?

Berry: We were instructed to crawl on the ground, as close as you could to the ground, because the overhead 'enemy' was shooting with blanks and some live ammo. So you were warned not to stand up or even rise up, but crawl and let the enemy shoot over you. If you stood, you might be hit by an actual live round. One guy from my hometown did actually rise up by accident and was hit by live ammo and killed. He never did make it out of boot camp to 'see action.' I did not know him nor knew about the incident until days after it happened.

Interviewer: What was the Army like during wartime?

Berry: At that time it was integrated, all blacks, at least in my quartermaster unit. But the officers were all white. My unit was in charge of service and supplies, transportation, food, clothing, laundry and providing recreation for the GIs and also to remove the injured from the front lines.

Interviewer: What were the most difficult things you dealt with in World War II?

Berry: Several difficult experiences come to mind right away. One is trying to dig a foxhole in the cold frozen ground under the cover of the Hürtgen Forest in Germany and during the Battle of the Bulge. Foxholes and trenches were dug with a standard military folding 'survival shovel.' It was very small because it could be carried easily in our backpack. We had to dig fast and dig deep to hide and be camouflaged from the enemy. Dig through frozen ground and frozen tree roots. In most cases, there was overhead fighting, but not direct fire. Digging foxholes was to prepare us for the approaching enemy.

It was also difficult to think about home—about you being without mom and dad, family and friends. But the idea was to have the enemy either surrender or be dead so the war would end so you could be home. That was your incentive to continue to move forward so you could get back home. To kill or be killed. Hurry up and let's get this war over with so I can go home!

Interviewer: How was the weather in the winter of 1944-1945?

Berry: The climate was difficult on my body because of the low temperature every day—the snow, the frozen ground. It was very, very difficult trying to get the blood circulating in your fingers. Gloves did not do any good. The temps were brutal. It's funny because when I was younger back home, below freezing temperatures were nothing to us. We'd be out all day sledding. We didn't mind the cold, but being in war, it's completely different.

Interviewer: How did you prepare for the cold?

Berry: A vivid memory is that the Germans had white uniforms and equipment. They were prepared for the snow. They had white camouflage uniforms. Their vehicles, tanks, trucks, etc. were painted all white to blend in with the snowy terrain. The Americans were not prepared for that region though because we had olive-drab green uniforms, trucks and equipment. We had to get orders back to headquarters fast and order white paint to repaint our vehicles from olive drab green, which stood out like a sore thumb so we could have a fighting chance.

Even though the British and Americans did not have white camouflage uniforms and equipment, we were still ordered to stop the enemy at all cost. Even in spite of our olive green uniforms and equipment that stood out like a fly in a glass of milk. We wanted to wait for the white paint but we still had to fight with our green equipment. We can't tell the enemy, 'hey, stop the war for a while so we can paint our equipment.' Somehow, we had to get it done while still fighting on the front lines in the frigid temperatures.

We never got our white uniforms. I'm not sure if the white soldiers got the proper color uniforms before the blacks, but I suspect they did.

Interviewer: What kind of weapons did you use during the war?

Berry: We had the smallest carbine rifle that Uncle Sam had, and we used it for target practice during training. We kept rifles but us blacks had to give back our ammunition. All during the war, we had no ammunition, until we had to fight in the Battle of the Bulge. But even so, we still had to keep our rifles clean and inspected. The purpose was probably due to a racial thing. Only a few us of had some ammo but only two or three bullets. I was one who had it.

We'd do target practice in the garbage dump piles. Rats were a good target. When we'd leave an area, we'd dump refuse in a hole that we dug and cover it up. We also dug big trenches for our outhouse.

Interviewer: Did you have any interaction with the local people?

Berry: The civilians would hear about us leaving an area and how we'd leave surplus supplies like soap, food scraps, etc. and they'd come soon after we'd leave an area and dig up things to use for their own families. When our GIs heard about the women washing in the streams and needing soap and fighting over our soap scraps, the GIs purposely let soap bars float downstream for the women to get, and not bury the leftover soap along with the garbage and other debris.

Interviewer: Did you see Holocaust survivors?

Berry: A vivid memory were the Jewish Holocaust victims, who were rescued by the Allies and brought into our camps, and how humanity could do those horrific deeds to other humans…just for greed. Even to this day, there are some flashbacks that remind me of those photographic memories and physical [strain] I went through.

Interviewer: Did the war change you?

Berry: It pretty much made me realize that you're not alone. You do have buddies, friends, and enemies. What I realize now is that all men are created equal but are not treated equal. Someone always wants to be on top and have someone else underneath of them. Nobody wants to be on the bottom. And you should not be [on the bottom], so try to reach always upward...higher. In this day and age, with technology, you can make it. It reinforced what I felt about humanity all along even before going into the war.

Interviewer: What was your most rewarding experience in World War II?

Berry: That I survived! It taught me to kill or be killed. Naturally, you want to shoot the first shot...and don't miss!

Interviewer: Did you have a diary or take photos to record your service?

Berry: I did have one camera and took photos in Luxembourg. Took my camera to a civilian photographer to be developed, but I got shipped out before I could get my photos back. I was very disappointed because one photo was of the 'Tiger Tooth' aka Siegfried Line. It was zigzag with spikes of iron from France—and throughout Belgium and other areas. It was an area that Hitler set up to keep the Allies out.

Interviewer: What was an event you remember when you returned home?

Berry: A prisoner of war was from the Crestmont section of Willow Grove, Pennsylvania, my hometown, who returned. I don't know which Japanese island he was on though. After the war, he was released and came home and was honored by our hometown by being in a parade. He rode on a fire truck as a hero, which wasn't done for blacks at that time. His last name was Upshore, but I don't recall the first name. He was younger than I. I knew his brother Jimmy and sister Hazel.

Interviewer: What do you want younger people to know about World War II?

Berry: The younger folks should look at WWII not as a history book back in 1944 and 1945, but also as a history lesson that lives. Human beings were not lost, but were taken due to greed of other countries who wanted to succeed in ruling the world.

This world was created for all. We started out as one human race and split off to various races throughout the world. Today, let's enjoy what freedoms we have and love your fellow man, putting God first: the love of God.

Since all men were created equal, it's only greed and being superior that's gotten this world in the condition that we're in today.

S. JOSEPH GIONFRIDDO: B-26 GUNNER FLEW DURING THE LARGEST AIRBORNE OPERATION OF THE WAR

Operation Varsity, March 24, 1945, was the largest airborne operation of World War II, involving nearly 15,000 airborne troopers of the U.S. 17th and the British 6th airborne divisions.

Supporting this airborne operation was the monumental task assigned to the Ninth Air Force. Less well known than the Eighth Air Force that flew the heavy bombers, B-17s and B-24s, on long-range missions over the Reich, the Ninth by 1944 was composed of the IX Bomber, Fighter and Troop Carrier Commands, numbering some 200,000 personnel, flying in support of the advancing American Armies.

During Operation Varsity, the Ninth flew thousands of missions to bomb enemy anti-aircraft positions, roads, and bridges in an effort to neutralize and isolate the battlefield. Joe Gionfriddo's outfit, the 397th

Bombardment Group, 597th Bombardment Squadron, flying twin-engine B-26 (Marauder) medium bombers, participated in that effort.

Joseph Gionfriddo in World War II.

After moving from bases in England to France in August 1944, the 397th had struck enemy positions in the Saint-Malo and Brest areas and bombed targets around Rouen as the Allied armies swept across the

Seine and advanced to the Siegfried Line. The unit began flying missions into Germany in September 1944, attacking bridges, defended areas, and storage depots. The 397th struck at the enemy's communications during the Battle of the Bulge (December 1944 to January 1945) and received a Distinguished Unit Citation for a mission on Dec. 23, 1944, when the group withstood heavy flak and fighter attack to sever a railway bridge at Eller, a critical link in German supply lines across the Moselle River.

We interviewed S. Joseph Gionfriddo, who, as a B-26 armorer/gunner, flew through some of that flak in support of Operation Varsity. Gionfriddo, who was 83 when we talked with him in Washington, D.C., kept a diary during the war to aid his memory.

Gionfriddo, who was 93 when he passed away in 2017, was named a Chevalier of the Légion d'Honneur by France.

Interviewer: Why did you join the Army Air Forces?

Gionfriddo: I was drafted in 1943. Initially, I was assigned to an ASTP, a special engineering group. I was sent to Lafayette College as part of the program, but unfortunately, they cancelled it. I then applied for pilot training, but during a physical, they told me I was color-blind. I ended up going to gunnery training.

Interviewer: Where did you train?

Gionfriddo: We started our armorer gunner training at Lowry Field in Denver. We spend about four months there. There are different types of armorer gunners, including radio operators and engineers. We didn't fly at Lowry. I remember the people in Denver were friendly to the military. After that training, we were sent to Barksdale Field, La.

We trained with French airmen at Barksdale, flying B-26s at the same time. We did fly out of Fort Myers, Fla., on B-17s and going through gunnery training. We flew all over the country. We then went to Hunter Field in Georgia for more training.

Interviewer: When did you get over to Europe? Where were you assigned?

Gionfriddo: After our training at Hunter Field, we had to go by British ship to Scotland in October 1944. We only spent two weeks in England, and then flew over to France in C-47s. By that time in the war, B-26 aircrews had suffered heavy casualties. We flew 30 combat missions out of the base at Péronne, France (home of the 397th Bomb Group). We were escorted on our missions by British Spitfires and P-51 Mustangs. However, the only time we saw RAF bombers were when 20 RAF bombers had to land at our base in France when they couldn't get back to England.

Interviewer: The Germans had a pretty extensive anti-aircraft system?

Gionfriddo: The flak was unbelievable. You could hear it, smell it and see it. On one mission a guy was wounded and said he was hit. I went to help him and noticed the metal parachute harness buckle had taken most of the hit. I liked being the tail gunner, though I had been a gunner on the top turret. I preferred the tail because you could see better back there and could get out of the plane very fast—something you always thought about.I manned twin .50 caliber machine guns. The only time I shot the guns was when they ordered us to shoot at a church the Germans were using as an observation post.

Interviewer: How was the weather?

Gionfriddo: The winter of 1944 was very cold. We flew with open windows at altitude in the bitter cold. The weather was so bad that when we taxied out on the runway, we got stuck in the snow and couldn't get out.

Interviewer: Did you know you were going to be part of Operation Varsity, the largest airborne operation of the war?

Gionfriddo: We had a briefing one morning where they said that all crews would have to go directly to our aircraft afterwards. They had MPs in the shack (to provide security) during the briefing. I did not know it was going to be the biggest airborne operation of the war—I was 20 years old. It was apparent that this mission was going to be something special.

We flew about 10 or so missions in support of the operations. Our pilot was Lt. Arthur Jessee, who now lives in Florida. Very thick flak. Our mission was to knock out 16 gun positions in a wooded area. Eight bomb groups were over the target before the 397th. We were the last group over the target. We finally were ordered to leave the area so the paratroopers could drop. This is around the same time as the operations around the Remagen Bridge, which we were warned to not fly around [because of the anti-aircraft batteries].

Interviewer: How was the Martin B-26 Marauder medium bomber? It did have a reputation as "The Widowmaker."

Gionfriddo: The B-26 was a good aircraft. I never liked B-25 (Mitchell). I also flew on B-17s, which is a nice aircraft, but not as fast as the B-26. B-26 carried same load as a B-17. We did have a crash landing in France when a B-26 hit a utility pole with a full load of bombs—two of the crew were killed in the crash. The original wing of the B-26 was pretty straight and hard to get off of the ground—took a pretty good pilot. Even [then Sen.] Harry Truman investigated the B-26 after a series of crashes. It took Gen. Jimmy Doolittle to intervene [to save the aircraft]."

Interviewer: Where were you on VE Day?

Gionfriddo: As the war grew to a close, we moved closer to the German border to Holland. We never flew missions there. I was there on VE Day."

Interviewer: What did you do after the war?

Gionfriddo: I came out and went straight to Penn State to get my bachelor's and master's degrees. I spent five-and-a-half years in the civil service and also worked for Army Intelligence. I also worked at the State Department marketing and selling strategic materials such as copper, zinc, and magnesium. President Johnson realized we could raise money this way—and in one year we sold $2 billion worth of materials.

I got involved in a project where the French built a road in Cambodia in the 1950s using laborers from Portugal. Also part of my job was helping to aid the Hungarian people after the 1956 uprising [against the Soviet Union]. We were involved in quick emergency funding through the Defense Department that ended up in Vienna. Other projects were in Central America when the United States was pushing for a unification of currency there. We also worked on many projects in Brazil.

Interviewer: Do you ever get back to your old base, or keep in touch with war friends?

Gionfriddo: I haven't been back to the base. However, several of my friends have. We get together through a B-26 veterans' group in Washington. One man lives near me and was in the same bomb group. We were on a mission over Frankfurt, Germany, in late 1944. One of the B-26s shot foil out of an open window to counter enemy radar. The plane was shot down—right next to mine. Our pilot asked us to count the chutes. We found out later that German civilians killed the bombardier when he landed. Six months ago, I went to the pilot of the airplane's funeral and talked to his son. He named his son after the bombardier.

Another member of the group lost a leg on a mission late in the war. When he went to the hospital, group staff officers showed him a book of airplanes to ask what type shot him down. He said that a (Messerschmitt) Me 262 jet aircraft shot him down and it happened very fast.

SGT HUGH MCGINTY: B-17 TAIL GUNNER IN THE "MIGHTY EIGHTH"

Hugh McGinty and his brother lied about their ages in 1941 in order to enlist in the U.S. Army Air Forces. After undergoing training at various bases, he qualified as a tail gunner on the B-17 "Flying Fortress." Later, as part of the ten-man crew of the "Blue Blazing Blizzard," 379th Bomb Group, Eighth Air Force, Sgt. McGinty found himself stationed at USAF Station 117, Kimbolton, England. They flew long-range missions over Nazi occupied Europe.

The U.S. Eighth Air Force was activated at Savannah, Georgia at the end of January 1942. In August conducted its first bombing mission, but not until a year later, when Roosevelt and Churchill agreed to Operation Pointblank at the Casablanca Conference was it given the mission that would give it lasting fame as "The Mighty Eighth," the strategic bombardment of the Nazi Reich. Before the Eighth stood down at the end of April 1945, its air-crews and fighter pilots had: flown 330,532 sorties; dropped 732,231 tons of bombs; fired 99,256,341 rounds of machinegun ammo; destroyed 15,731 enemy aircraft; and suffered 46,456 casualties. (The names of those men, along with casualties suffered by the RAF, are hand-inscribed in Books of Remembrance kept in St Clement Danes, a small 1682 Christopher Wren church located in London's Strand. The church was bombed by the Luftwaffe in 1941, rebuilt and dedicated to the RAF in 1958.

Rob Woodard spoke a few years ago with McGinty, who passed away in 2017 at 98.

Interviewer: What made you want to join the Army Air Corps?

McGinty: Well, I grew up in the Philadelphia area and always liked adventure and had been interested in aviation. So, in January 1941, my brother and I enlisted.

Interviewer: You enlisted even before the United States had entered the conflict, why was that?

McGinty: The depression was on, and I had lost my job and I didn't want to depend on my parents, I wanted to get out. Plus, I had an uncle who had served.

Interviewer: Where did you go to basic training?

McGinty: Lowry Field (in Denver), but it wasn't boot camp, I had eight days of recruit training and went right on the flight line. I worked as a mechanic."

Interviewer: How did you make the transition from mechanic to gunner?

McGinty: I was up in Utah and was on the firing range one day and an officer came up to me and said, 'Hey, you're pretty good, you want to go to gunnery school?' He really sold me on it, said I would go up in rank much faster. So, I decided to give it a try and was sent to Las Vegas for gunnery school. Then to armament school back at Lowry again, then Will Rogers Field in Oklahoma to train on light bombers and then from there I went to Salt Lake City where they assigned me to a B-17 outfit. Finally, I went to Texas for training as a tail gunner. After finishing up training in Texas they shipped us over to England in 1943.

Interviewer: So how much actual time did you spend in England?

McGinty: I got there in November of 1943 and flew until May 1944.

Hugh McGinty served in famous unit.

Interviewer: The army credits you with 36 missions, where were most of your targets?

McGinty: I only credit myself with 29. The Army must have counted some missions that were called back. Our targets were all over

Germany—Berlin, Schweinfurt. We went to Poland, some targets in France, and Bremen, Germany.

Interviewer: As a gunner, you're charged with protecting the bomber from enemy fighters, did you encounter them on a pretty regular basis?

McGinty: Oh, yeah, all the time. We were kind of thinking that we weren't going to make it most of the time. Our losses were pretty high. Our chances of survival were about one in five. But I think our main problem was fatigue most of the time. You fly 10-hour missions with oxygen sometimes day after day, you get so tired. You get up at 3 a.m., maybe you don't fly until 6 p.m., but you've got all the preparation, then you get back, you've got to clean the guns, eat, and try to get some sleep with all the stuff going on in your mind. I think fatigue was one of the worst things I encountered.

Interviewer: How was it decided who would go out and who would stay in?

McGinty: The crews were assigned to go, 10-man crew, and they would sort of rotate the crews, sometimes they'd put 21 planes up, sometimes they'd put 40 airplanes up, depending on how many planes they had and what the mission was. But they'd rotate the crews, say you had 21 flying, they'd send three spares that would go and then they would come back if nobody aborted.

Interviewer: Is there a particularly harrowing experience that stands out in your mind, where you thought, 'this is it?'

McGinty: Several. On the 29th of February we flew to Frankfurt, Germany. Just as we were approaching the target, we had a nose attack, you know, fighters attack at the nose. I heard the nose gunners calling these fighters out and I hear the bombardier say the navigator is down. And then I heard the pilot call the top turret gunner and tell somebody to get him an oxygen bottle because our oxygen's been shot out. And the turret guy says, 'I'll get you one as soon as the fighters go by.' And the pilot says, 'The hell with the fighters, get me some oxygen, I'm passing out!' Next thing I know, we were in a spiral going down, the airplane just left formation and started down. I called the nose and didn't get an answer.

Interviewer: Did you have to bail out?

McGinty: I decided, from experience, that you've got to make a decision in a hurry to get out of the airplane before it goes down or explodes, breaks up, whatever. So I unplugged my communications and heated suit, and instead of wearing a parachute, I would wear a harness, and I used to leave my parachute next to the escape hatch, but when I got there it wasn't there. It had slipped up forward past the tail wheel into the waist gunner's position.

The centrifugal force of the airplane wouldn't let me get to it, so I just laid back down on the catwalk and thought, 'this is it, I'm not going to bail out without a chute.' I put my feet forward so maybe I would live a second longer and also not to bust up a pretty face. Then all of a sudden the centrifugal force stopped because we went into a straight dive. As I moved toward my parachute, I could see the gunners up in the nose, they were talking to somebody on the intercom. I went back and hooked up my communications and I heard the pilot say, "Okay, don't bail out yet."

Interviewer: Why did that happen? Did you ride the plane in?

McGinty: [The pilot] regained consciousness about 15,000 feet. We started (falling) at 27,000. Between him and the co-pilot and the engineer pulling back on the yoke, they got the airplane out of the dive. The bombardier had dropped the bombs, making the plane lighter; otherwise they would have probably pulled the wings right off the airplane. We get back to level flight at 4,700 feet then we pulled up to about 10,000 feet, pulled up into some clouds. I get back in the tail and I see three fighters sitting right on top of the clouds, our tail is like a shark's fin sticking out.

I yelled to the pilot, get the thing down, these fighters were getting ready to fire, just as we started down into the clouds 20 mm fire was flying all around us. They look like little sparklers. I don't think that they thought it was worth it to risk a collision in the clouds to come after us. When we got over France it was clear, some light arms fire and a little flack; then we crossed the English Channel to go home. Back over England, we landed at an RAF base and then we took the navigator out. He had charted us a course for home. He had been shot in the stomach, but before he died, he plotted us several courses from various points.

Interviewer: Quite a story.

McGinty: Pretty harrowing.

Interviewer: What was life in England like, were you permitted to leave the base much, what did you do there when you weren't flying?

McGinty: Oh, the English were so great. Well, we'd go into a restaurant and everything was rationed over there but they'd have a little meat set aside. You didn't know what kind of meat it was of course. Their clothes looked a little outdated to ours, because everything went to the war effort, they wore cotton stockings, not nylon. You know, little differences.

Interviewer: How about around the base, if you weren't on or preparing for a mission what would you be doing?

McGinty: Sometimes, you would work with the ground crews, help them with the planes. I had been a mechanic so I could do that. We had support teams, we had movies.

Interviewer: Did you fly with the same crew the entire time you were there?

McGinty: Most of the time, yes.

Interviewer: I read that the King and Queen of England visited Kimbolton, were you there for that?

McGinty: Yes, actually what happened was pretty funny, we had a plane that the crew had named '4Q2,' which we weren't sure the royals would appreciate so before they came the crew covered it up with 'Princess Lizzy' or something like that. And she came by the airplane and it was kind of dedicated to her, she shook hands with the crew, all of that. When they left, the crew wanted to put the '4Q2' back, they were superstitious about changing the names of the planes, but somebody said no, it's obscene, so they just painted 'censored' where the old name was.

Interviewer: Are you still in touch with any of your old crewmembers?

McGinty: All but one are dead.

Interviewer: Do you get together at all? Do you reunite with members of the bomber group?

McGinty: We have reunions of the bomber group. The last one was in Seattle.

Interviewer: Have you been back to Kimbolton?

McGinty: I went back in 1991, when I went to Germany and England. I was going to go back to the base and I stopped at the cemetery at Madingley, near Cambridge [Cambridge American Cemetery and Memorial]. My navigator was buried there. When I saw all of those crosses…It is a beautiful cemetery now, but back then it was a mud hole. I went up to Bedford, which was near the base and I started to go to the base, but I just couldn't. Emotionally, I just couldn't. I broke down, so I went to London instead.

Interviewer: What did you do after the war?

McGinty: I stayed in the service. Made a career of it and stayed in until 1962. I got out for 10 months or a year and then I re-enlisted at Lowry Field.

Interviewer: What were you doing?

McGinty: I went on the Berlin Airlift in '48. Came back from that and then they sent me to Texas. From there, they shipped me to Goose Bay, Labrador, and I spent 32 months in the Arctic in a rescue unit.

Interviewer: Anything else you think our readers should know?

McGinty: Well, it's not all fun and games, it wasn't fun. You lost a lot of friends. You got so you wouldn't make friends anymore. When the new crews would come in you would (not associate with them). You would almost hope that if there was somebody less trained and going to get killed, it would be them. So, you didn't make friends with them. It got to be very clannish with your own crew--you got to be like brothers. When you lost someone, it was like losing your family. You depended so much on these men, so there was no room for error. If you made an error it was deadly. Each man knew what part of the airplane he was supposed to protect, we didn't try to be aces, and we tried to keep the enemy off of us, discourage the hell out of him so he'd look for an easier target. The main thing was survive, survive, survive. And we did…except for one man.

DON WHIPPLE: WOUNDED IWO JIMA MARINE SAW FLAG RAISED

Don Whipple grew up on a farm in western Kansas, never dreaming he would play a role in one of the most famous battles in United States history.

Whipple, 95 and living in the Denver area when interviewed, grew up during the Depression as a member of a large family. He attended a one-room school house.

He persuaded his parents to sign the papers for him to enlist in the Marines—as he was not yet 18 years old and right out of high school. He joined the new 5th Marine Division, which was made up of both combat veterans and new recruits. Some of the veterans were from the 1st Marine Parachute Regiment, Raider Training Battalion, parachute training schools, and replacement companies.

The division moved to Hawaii to continue training at Camp Tarawa. At Christmas, while training in Hawaii, Whipple learned, by telegram, that his father had been killed in a car accident in August

1944. It was a devastating moment for Whipple, who sailed with the division to fight on Iwo Jima a few months later.

The 5th Marine Division fought on Iwo Jima from its initial landing on 19 February to 26 March, 1945, before returning to Hawaii. The initial landings were southeast of Mount Suribachi.

The division paid a terrible price on Iwo Jima, with 2,482 killed in action, 19 missing and 6,218 wounded, including Don Whipple. It was the highest casualty rate among the three attacking Marine divisions.

Don Whipple experienced the bitter fighting on Iwo Jima.

On Iwo Jima, the division earned 17 Medals of Honor (15 Marines and two Navy corpsmen). Twelve of these medals were awarded posthumously. In addition, the unit was awarded the Presidential Unit Citation as part of the Fifth Amphibious Corps.

The 5th Marine Division was to be part of the invasion of Japan under Gen. Douglas MacArthur and the 10th Army, but Japan surrendered after the dropping of the atomic bombs. After the surrender, the 5th Marine Division landed at Sasebo. Whipple and his unit had occupation duty in Nagasaki, which recently was leveled by the atomic bomb.

To this day, Whipple carries Iwo Jima with him. Literally. Six months before this interview, he rubbed his knee joint and a small sliver of steel shrapnel fell out into his hand. In 1953, he had an operation to take another ragged piece of shrapnel out of his leg. "People wanted me to keep it as a souvenir, but I just threw it away," Whipple said.

Whipple, a member of Denver's Veterans of Foreign Wars Post No. 1, frequently speaks to high schools about his experiences on Iwo Jima to let students understand the importance of not forgetting the sacrifices made there. He also assists younger veterans who served in Afghanistan and Iraq.

Whipple was honored, with other Colorado Iwo Jima veterans, by President Trump and Vice President Mike Pence when they came to the state in early 2020.

Interviewer: When did you decide to join the Marines?

Whipple: I had three older brothers in the Navy. I had read books like *They Were Expendable*, which was interesting. I saw a picture on the cover of the *Saturday Evening Post* about the Marines at Guadalcanal. It seems that was where the excitement was. I graduated from high school, then boarded a train the next day to Denver to enlist in the Marines.

I returned to Kansas and worked on the farm because the Marines didn't call me up until August. My brother and I were the same size and joined at the hip. We were very competitive. We would break horses for ranchers to ride and that gave us spending money. We were the last of the real American cowboys.

Interviewer: Where did you train and for how long?

Whipple: When we got into San Diego for training at midnight, the whole area at the Marine base had camouflage netting. A Marine came and told us, 'Welcome to San Diego Marine Base, if you want to know where the head is, it is over there,' and he pointed. I had no idea what a head was. I thought was the commander, but found out it was the toilet. We were assigned huts, marching everywhere initially in civilian clothes. I remembered seeing a lot of older Marines at the Marine Depot, with their rifles stacked, asking us where we were shipping out to and then telling us, 'you'll be sorry.'

I had no idea how tough the training was going to be. I got through it in good shape. The 30-foot drop into a pool was a challenge to me. After boot camp, I went to Communications School where I learned radio and telephones and to lay wire. At Camp Pendleton, I also became an artillery forward observer. I was assigned to the 13th Marines, which is an artillery unit. We had veterans of Guadalcanal and other campaigns with us.

We trained to go over the side of a ship into LCVPs [Landing craft, vehicle, personnel]. We would load up and do it again. Some of our exercises were out on Coronado Island. One day we were told, when big limousines rolled up, that President Roosevelt was there. We practiced an amphibious landing. We had a final inspection where our packs and other equipment had to be ready.

We boarded troop ships with no idea where we were going. We sailed to Hawaii. This is the first time I had ever been on the ocean. Even most of the lakes in Kansas are not natural."

Interviewer: What did you do in Hawaii?

Whipple: Six-to-eight months were spent on the big island of Hawaii training. We trained there at Camp Tarawa, on the far end of Hilo. There was a lot of volcanic sand and dirt. It was the nearest thing to Iwo Jima, which they call Sulphur Island.

It was just exciting time for us teenagers. Years later, President Reagan said that World War II was won by a bunch of teenagers.

Interviewer: When did you know you were going to Iwo Jima?

Whipple: We shipped out of Hawaii, stopped at Eniwetok and wound up on Saipan. We finally loaded up for Iwo Jima. It was a huge convoy. On the troop ship, they had maps of Iwo Jima. I had never heard of Iwo Jima. They told us it was going to be a rugged time, and guys were not coming back, but it would only be eight-to-10 days of fighting. We knew that it was halfway between Tokyo and Guam and vitally important to our air crews. They had bombarded the island with B-17s and B-29s prior to our arrival. The Navy's battleships and cruisers fired broadside after broadside into the island. An old Kansas hayseed like myself was very impressed.

Interviewer: What happened on the initial landing?

Whipple: Onboard the troop ship, we were up at 3 a.m. for a dinner before the invasion. It was a breakfast holiday dinner with turkey, dressing, and pumpkin pie. Then we went topside to see the Navy firing at the Island. At 6 a.m., we went over the side and got in line with the other landing craft. We had done it so many times in training that it was routine. There were 27 of us in the landing craft.

During the bombardment, it looked like the whole island was on fire. We joined other landing craft going in circles. We were in the second wave. The night before, I had eaten a greasy meal, so I heaved—fed the fishes. It is the only time I have ever been seasick.

We saw and heard a Navy fighter plane, a spotter plane for the battleships, sputtering, then the motor would kick back in. Then it sputtered again and dived with the nose down and started to spiral. The Navy pilot was off one of the aircraft carriers. We listened to the radio and learned that he had received a telegram that his wife gave birth to an eight-pound baby. We heard that he was singing as his plane spiraled and splashed in. He never came up. Again, we thought these guys are out to kill us—this is really a war. The war became real to me. This poor guy would never see his wife and baby.

There were splashing bombs and a lot of flames. A landing boat to our right was hit from an enemy shore gun on Mount Suribachi and went straight up in the air. Men were falling out of it. I again realized that these guys were out to kill you. When we landed, we hit the deck into the black ash and tried to shrink down. We saw machine gun tracers firing over us and took light artillery and mortar fire. Captain Austin told us to get off the beach and follow him. Wire and rope was used to pull machine guns up steep hill slopes. The Japanese had made terraces in the beaches, with bulldozers, which were hard to walk on.

You would sink to your knees, then slide back another two feet—it was like quicksand.

Interviewer: What happened when you were wounded?

Whipple: I was ahead of a few people, then stood up to see what was on the other side of an ash hill and turned around. All of a sudden, a mortar round came down and I saw a black bullet-shaped shell skip five feet away. There was a tremendous blast, then light, then everything went dark. I was hit by shrapnel.

I laid there a few minutes. Captain Austin thought I was dead. I tried to get up, but he told me I had been hit and to lay there and wait for a [Navy] corpsman. I didn't want to be taken out [of the fighting] or put other guys at risk who were trying to help me. But the captain demanded that I be taken to an aid station, where they were triaging Marines on the beach. I would come to and then pass out. The beach was littered with guys with all types of wounds. They took me aboard an LST [landing ship, tank], then I was taken out to a troop ship that was made a temporary hospital ship.

Interviewer: What happened on the troop/hospital ship?

Whipple: I had sustained a head injury and had shrapnel in my knee. I was taken by stretcher to a bunk and noticed the guy next to me was in bad shape, with artificial skin that looked like black plastic. There was no flesh from his head to waist. When he would try to talk, he would scream. Another initiation to war.

The troop ship was emptied, all boats were gone. The Navy has bunks several rows high. I saw a kid, who looked like he was only old enough to have been a junior high student. He seemed doped up and collapsed on a bunk. The next morning, the ship's captain came in and said they were having a burial at sea. Many had died overnight, including the young kid. They had a chaplain, [a detail of] sailors who gave a rifle salute and then they flipped the dead off of paddles into the sea. It was a sad experience and I cried like a baby when this took place. These kids never had a chance at life and died in one of the loneliest places to die.

Interviewer: Was the war over for you?

Whipple: After [the ceremony] was over, standing in front of the line, I recognized the voice of another forward observer, Richard Thomas, who got hit in the back. We both agreed we had to get off the ship and back on the island. We went to the bow of ship and saw a sailor, a coxswain, preparing a boat. We asked if he was going to shore and he said not until the next morning. We asked him if we could get a ride and he said yes, but we had to be there at 8 a.m. the next morning. We had Navy pajamas on, no helmet or rifle, no dungarees, nothing. We saw a big pile of dead men's gear that had rifles, helmets, boots, clothes, so we outfitted ourselves. We went back to our bunks. The next morning we went to the boat and were helped into it. They let the boat down with a winch.

Interviewer: Did you rejoin your unit? How did you find it?

Whipple: Our boat landed on Green Beach, at the foot of Mount Suribachi. I was hobbling on one leg as it had been hit and had a head injury. The Marines had been taking a beating. We didn't know if we would have anyone left in our outfit. We found the communications base near an old truck with unit markings on it. The commo center had a tarp over it, with sandbags on top of it. When I arrived, one guy said, 'Whip, we thought you were dead.' Which I replied, "no, not yet."

Interviewer: Was there a leader on Iwo Jima who stood out?

Whipple: Our commander, Captain Robert Austin from Greenville, Tennessee. He was a Marine's Marine and was a Guadalcanal Marine Raider. He often held our unit reunions in his town. He is buried there in the town's cemetery. He hired men from my outfit after the war.

Interviewer: What was the mindset of the Japanese defenders?

Whipple: The Japanese general in charge once lived in America and knew about our way of life as he was posted to an embassy. He told his troops that they would never leave Iwo Jima alive and they were each responsible for killing 10 Marines before they died.

Interviewer: What was your job on Iwo Jima?

Whipple: I was a forward observer and had to help the guns calculate how to reach their targets. I would go forward and find a crack in the mountain and set up and see the battlefield. The [Fire Direction Center] would fire a round to establish a basepoint. They would fire at an old pillbox that the bombers had knocked out. I would see how far they missed it and tell them to adjust. Sometimes this would be 'range correct, four millimeters to the left, fire another round, etc.'

I directed fires from that crevice for 55 hours straight without relief. My leg healed up a bit. Finally, I felt I could walk and called the FDC and followed a rifle squad going up Mount Suribachi to an [observation post]. I would find an opening in a rock area, make a foxhole, and call the FDC that I was in place.

Interviewer: What were the conditions like on the island?

Whipple: We were in continuous combat. I did not have a bath for 36 days. We had no hope for anything. We had seen the worst brutality human beings can do. The smell of the dead, bloated bodies knocked you over. A small vehicle with tracks had a dumpster bucket on back that was loaded with dead guys like cord wood. There was no drinking water. What you found was poison in water-filled cisterns. Sulphur in the sand smelled like rotten eggs. The skies were yellow. There were decaying bodies everywhere.

Interviewer: Did you see the famous flags raised on Mt. Suribachi?

Whipple: One group of guys with the flag wanted a hidden machine gun taken out so they could continue. It had been in a cave. I called the FDC, so they fired a round and overshot, but then fired the next in the center [to knock out the machine gun]. I was in that forward spot for days. Those kids walking up and down [Mount Suribachi] went through hell.

Some guys were looking for something to put the flag on. I think they found a long pipe. I watched [the first flag] go up. I actually saw both go up. It was amazing and a real honor to watch history happen.

Interviewer: How did the battle end for you? Was it worth it?

Whipple: The last night before they surrendered, the Japanese launched a banzai attack. It was pure suicide and a waste of life.

There were thousands of airmen saved by landing on Iwo Jima. They would have had to land in the ocean, or be captured by the Japanese. They were cruel to prisoners of war. Many of the B-29s were shot up and forced to make crash landings on Iwo Jima. One year, we had a reunion in Tucson and wondered why an Army Air Corps group co-located their get-together at the same time and place. We learned that they wanted to thank us as many of their air crews were saved by landing at Iwo Jima. We all got together with our families and there wasn't a dry eye in the house that night.

Interviewer: There has been much discussion, since those days, as to whether or not the United States should have bombed Japan with atomic weapons. What are your thoughts?

Whipple: They were saying there would be more than a million casualties, and millions of Japanese lives lost as well, in an invasion of Japan. The bomb saved a lot of lives. When we came back to Hawaii to refit, we were halfway across the ocean when they told us that Japan had surrendered. We were so beat up that they sent us back to Hawaii to get replacements. We knew we were going to be in the invasion of Japan.

Interviewer: You were ultimately part of the occupation forces after Japan's surrender?

Whipple: We landed at Sasebo. They made us MPs for a day or two. We noticed when the Japanese surrendered, they really surrendered. We took all of their arms and ammunition that were in caves and destroyed them, burning them in the streets of Nagasaki. We patrolled in the epicenter of the blast. There was fine dust that we didn't know about. There was a destroyed kamikaze base on the north side of town.

Kids would come out to the curb in Nagasaki and we would give candy to them and they would sit on our laps. Mothers would sing folk songs on their side of the street and we would sing back.

Interviewer: What did you do after the war?

Whipple: My wife Joann and I were missionaries for 15 years. We had a daughter, Debra, and lost a son at childbirth, who I hope to meet in heaven. I knew I was a sinner. I prayed in those rocks on Iwo Jima and asked Jesus to come into my heart. I wanted to know how I could appropriate what Jesus did to be a good person. God has been a blessing to me.

DONALD BEATY: USS INDIANAPOLIS SURVIVOR

Many people's only knowledge of the USS *Indianapolis* comes from a few lines uttered by the grizzled character Quint, played by Robert Shaw, in the 1975 movie "Jaws." However, hundreds of thousands of lives were linked to the ship in the closing days of World War II.

The heavy cruiser USS *Indianapolis*, commissioned in 1932, earned 10 battle stars in the Pacific. She had bombarded many Japanese-held islands including the Aleutians, Tarawa, Saipan, Iwo Jima and Okinawa. After shooting down several Japanese airplanes, *Indianapolis* was hit by a kamikaze off of Okinawa with the loss of 13 of her crew.

The cruiser was then ordered to Mare Island Naval Shipyard for refitting and while at Mare Island was ordered to carry "Little Boy," the first two atomic bombs, to the island of Tinian in the Marianas. After delivering the bomb parts, pausing at Guam to shift personnel and then steaming for Leyte, the Japanese submarine I-58 intercepted and then sank the cruiser with two well-placed torpedoes. The top heavy *Indianapolis* rolled over and went down by the bow in 12 minutes.

Of the 1,196 sailors onboard *Indianapolis* when she sank, about 900 survived; of those, only 316 survived the next four days while adrift in the Philippine Sea. Almost 600 sailors had died from injuries, drowning, delirium, dehydration, and shark attacks. Soon after a patrol aircraft spotted those men on the fourth day of their ordeal, a PBY floatplane arrived on the scene, dropped life rafts (unfortunately too far from the exhausted survivors to be of much help) and then, defying standing orders, landed in the twelve foot seas to aid many of the survivors. Later, the plane, unable to become airborne, was destroyed by naval gunfire.

The skipper of *Indianapolis*, Capt. Charles Butler McVay, USN, was eventually court-martialed for the tragedy on two counts and found guilty of one—"hazarding his ship by failing to zigzag." Later, Adm. Chester Nimitz, who disagreed with verdict, returned him to active duty, allowing McVay to retire from the Navy as a Rear Admiral in 1949.

Hounded by the families of the lost seamen, the retired admiral took his life with his Navy-issue revolver in 1968.

In 2000, spurred by letters from 11-year-old Hunter Scott and members of Congress, the U.S. Navy finally cleared McVay's service record.

We interviewed Donald Beaty, then 80, who served on *Indianapolis* from 1943 to 1945, and was onboard during the devastating torpedo attack. Beaty, a retired industrial insulator, passed away at 87 in 2012.

Interviewer: Where did you join the Navy? What did you do in the service?

Beaty: I was drafted in 1943. I went to boot camp at Great Lakes, Ill., then reported directly to the ship. The ship was located at Mare Island in California. I was a Seaman First Class and a Fire Controlman.

Donald Beaty survived Japanese torpedo, sharks.

Interviewer: Tell us about the ship, are there any special memories?

Beaty: It was known as a fast ship. It had a stateroom built for President Roosevelt. It also was the flagship of Adm. Spruance, who was a hell of a commander. He didn't get that much publicity, but he was the man who won at Midway.

Interviewer: Was there word out that you had a secret mission coming up after Okinawa?

Beaty: At that time, we thought the war was winding down. Many of us thought the area we were in was secure.

Interviewer: Did you have any inkling of what your special cargo was?

Beaty: The whole crew talked about a 'secret weapon' onboard. It was on the hangar deck surrounded by Marine guards.

Interviewer: Where were you when the first torpedo hit?

Beaty: I had just gone on duty. I was manning a 40 millimeter (Bofors anti-aircraft gun), which was a new system. I knew something had happened in the boiler room. Our communications systems were knocked out.

Interviewer: What were the conditions like after abandon ship was ordered?

Beaty: All we had on were our kapok jackets to keep us afloat. We had no drinking water. We knew better than to drink sea water—you were dead if you did that. There was some pushing and shoving in the water after a few days. A few of us were on a floating net, others were on rafts. Our shoes were off. The dead attracted the sharks. The sad thing is that the Navy didn't come and get us.

Interviewer: How were you finally rescued?

Beatty: Planes went over us every day, but did not stop. [After a few days,] one airplane saw our oil slick and radioed it in.

Interviewer: Tell us about your captain, was he a scapegoat?

Beaty: There is no doubt about that. He was exonerated in a trial in Washington, D.C. The whole thing was ridiculous. That young boy [Hunter Scott] helped to clear his name many years later.

Interviewer: You've had a few years to think about the significance of the cargo the ship delivered, what are your thoughts?

Donald Beaty often spoke to students about his experiences.

Beaty: [President Harry] Truman is my hero. The Japanese warlords would not surrender. The dropping of the atomic bomb saved hundreds of thousands of lives. However, the destruction caused by the bomb surprised all of us.

Interviewer: What did you do after the war? Do you still see your shipmates?

Beaty: After the war, I was through with the ocean and came home to Indiana. I worked for a General Motors plant in Angola, Ind. I still see Jim O'Donnell, who lives in Indianapolis and was on the ship. We try to get together every two years down south (in Indianapolis). I also talk to school-children about the ship and the war. I try to explain what we were fighting for.

BILL BARNEY: B-29 RADAR OPERATOR OVER NAGASAKI

Much has been written recently about President Truman's decision to drop our two atomic bombs on Hiroshima and Nagasaki to end World War II. Of these missions, less is known about the second Nagasaki raid (August 9, 1945) that convinced Emperor Hirohito and the Japanese high command to capitulate. Our interviewee, Airman William Barney, flew on that mission in a B-29 nicknamed *The Great Artiste* (V-89) instead of in his usual *Bockscar*. The reason for the switch is interesting and has been caused confusion ever since.

After the Aug. 6, 1945, Hiroshima mission, Charles Sweeney, who was commander of *The Great Artiste*, was chosen to carry the atomic payload for the Nagasaki strike. On the Hiroshima mission, Sweeney and his crew had flown the *Artiste*, configured as the instrument plane. They trailed Col. Paul Tibbets' *Enola Gay* (V-82). Since *Artiste* was already set up as an instrument plane, it remained in that configuration for the Nagasaki mission, which resulted in Sweeney's crew swapping planes with Capt. Frederick C. Bock's B-29 crew. Bock and his crew flew the mission in Artiste, while Sweeney and his crew dropped the "Fat Man" atomic bomb from the *Bockscar* (V-82).

John Wassell, who serves as historian of the 509th Composite Group, says the airplanes had small numbers on their nose (with no nose art), so the planes often were reported incorrectly. Many believed that since Sweeney's crew dropped the bomb, it must have been from *The Great Artiste*, but that was incorrect.

Bill Barney witnessed atomic bomb dropping.

Wassell said he became involved with the 509th reunions to keep history alive - and to free it from revisionism. "My father, after having fought across Europe with the 12th Armored Division, was poised to be part of the invasion of Japan. Had it not been for the bombs, my father, and then of course myself, may not be here today."

Rob Woodard interviewed Bill Barney, then 82, the radar operator on the *Bockscar*, from his Columbia City, Ind. home, where he lived with his wife, Cora, before his death in 2016. Barney normally flew on the *Bockscar*, which was part of the 393rd Bombardment Squadron, 509th Composite Group, 20th Air Force, on all of its missions from Tinian, but, as previously explained, on the Nagasaki mission he flew in the Artiste.

Interviewer: You were stationed on Tinian, what was that experience like?

Barney: It was just an island. There was nothing there, just people, it was all GIs, but there were a bunch of Koreans there that the Japs used to do their work for them.

Interviewer: What was your participation in the bombing of Nagasaki?

Barney: On this particular day, I wasn't on the *Bockscar*, as it had the bomb. Maj. Sweeney was the pilot on that mission. We had Maj. Sweeney's plane, *The Great Artiste*, which was the plane that was set up for blast measurements, and set up with special equipment. So on that day, in addition to being radar operator on *The Great Artiste*, I was running one of those pieces of equipment that measured the blast. Col. Tibbets, who was the commanding officer, asked Sweeney, who was second in command, if he would like to fly the bomb on the second mission. Sweeney said, 'yes, but I would like Fred Bock's crew to go with me.' So, Tibbets said okay, that's simple, just switch planes with Bock for the day."

Interviewer: Sounds simple, so Bock and Sweeney switched planes. What sort of equipment was on *The Great Artiste*?

Barney: We had three canisters used to measure the blast effect. As soon as our bombardier saw the bomb come out of the lead plane, he dropped these three canisters.

Crew of the Bockscar.

Interviewer: They were dropped?

Barney: Yes. Each canister had a small parachute and when it was dropped, it drifted down. Each of them was about 14 inches long and eight inches across. As soon as they were dropped, we had these boxes on our plane that locked onto them and measured the blast—the power of it. Two of them were operated by scientists, and I was running the third one. I took measurements from one of the canisters.

Interviewer: Were you able to see the actual physical damage on the ground?

Barney: We were watching the [telemetry] machines--so by the time the blast happened, we were probably four, five, or six miles away. We were quite a ways away.

Interviewer: Could you see the mushroom cloud?

Barney: Oh, yes.

Interviewer: There has been much discussion, since those days, as to whether or not the United States should have bombed Japan with nuclear weapons. What are your thoughts?

Barney: I don't think there was anything to feel bad about dropping them. After all, they started this [war]. Everybody knew that if we had invaded Japan, there was going to be a lot of lives lost on both sides…and this way it was all Japanese lives. That's the way I look at it, I would rather have it that way.

Interviewer: How many men were in your crew?

Barney: We started out with 11, but after we got started on this special mission, they refitted the planes. They took our guns away. We only had guns in the tail, so after that, we had a crew of nine.

Interviewer: Do you stay in touch with any of them?

Barney: Oh yes, there are four of us left of our crew, our next reunion will be in October in Washington, DC. We have a reunion every year. There are getting to be fewer of us every year, of course, but second- and third-generation (families) are filling in to keep the stories alive.

Interviewer: Who are the four left?

Barney: On the 50th anniversary in 1995, all nine of us were still living. Now the guys left are Bob Stock, he was the tail gunner; Hugh Ferguson, co-pilot; Len Godfrey, he was the navigator; and myself.

Interviewer: After Japan surrendered, what did you do?

Barney: We stayed on Tinian, we didn't get leave over there until November. I got home on Thanksgiving Day, 1945.

Interviewer: What did you do after the war?

Barney: I was a farmer. I'm a retired farmer now.

ALFRED CONSIGLI: TANKER FOUGHT UNDER PATTON

Alfred Consigli, age 100 of Upton, Mass., was a tanker in the 774th "Blackcats" Tank Battalion that fought under Gen. George Patton during World War II. Consigli, who was drafted, landed at Utah Beach, saw heavy fighting in Normandy, and fought in the Battle of the Bulge. He was a U.S. Army tank gunner and crack sharpshooter.

Consigli grew up in Milton, Mass., and had seven brothers, out of nine total, who served during World War II (two were not old enough to serve). He said none of his brothers ever talked about the war.

Consigli went to basic training at Fort Knox, Kentucky, where he was designated a sharpshooter. He took a troop transport and trained in England to prepare for the eventual invasion of Europe.

After landing at Utah Beach, Consigli, who said he fired more than 8,000 tank rounds during the war, was in constant combat for nearly a year. His unit fought into Germany after crossing the Remagen Bridge.

Consigli next to his tank in Normandy, France, in 1944.

He saw the horrors of war, seeing dead men from both sides lying in ditches and in the fields. "There were so many dead, in the road and in deep trenches. We saw one soldier laying there with his wedding picture and you could tell he'd just gotten married," he told Linda Chuss of the *Milford, Upton & Mendon Free Press.* "Then a tracer came overhead and we quickly backed up our tank; the next shell would have hit us. They were going to promote me after that but soon the war was over."

Consigli has one story where he and his tank crewmembers slept in a barn to keep warm. He was astonished to learn that U.S. Army infantry had captured German soldiers who were sleeping in the same barn in the hay loft.

Alfred Consigli then and now

After the war, Consigli went home to Massachusetts – as did his six brothers who were in the service. In 2020, he told the Worcester, Mass., *Telegram & Gazette* that his mother's untimely passing at 43, weeks after he returned, was probably due to her worrying about her sons in the war.

Consigli wanted to remind others of the sacrifices that were made. He often gave talks at schools, veterans' organizations and other groups. In addition, he displayed his wartime photos at the Patton Homestead in Hamilton, Mass., during the annual military event each summer. He also gave a presentation about his tank service at the American Heritage Museum in Hudson, Mass.

MICHAEL W. GREEN: AMERICAN BATTLE MONUMENTS COMMISSION SUPERINTENDENT, CARTHAGE, TUNISIA

The bitter battle for North Africa was the U.S. Army's first taste of ground combat in the Mediterranean Theatre. Many of the fighting men who fell in that struggle are buried in a little-known American Battle Monuments Commission (ABMC) cemetery in Carthage, Tunisia.

Congress created the ABMC in the aftermath of WWI (1923) to create and administer American military cemeteries and memorials honoring the men and women of the American Expeditionary Forces. Gen. John J. Pershing served as Chairman of the Commission from 1923 to 1948. Under his direction, the ABMC created cemeteries and memorials that are impressive for their monumental size and beauty. Today, the Commission administers some 26 overseas cemeteries along with overseeing an additional 30 additional memorials, monuments and markers, almost all dating from WWI and WWII, in

17 foreign countries and the United States (3). ABMC headquarters are located in Arlington, Virginia and it maintains an office in Paris.

We interviewed Michael W. Green, then superintendent of the North Africa American Cemetery. Green, a former Marine who had experience in several of the U.S. military cemeteries in Europe, said that Americans are making their way to the North African cemetery in record numbers. Green, now retired, lives in Virginia Beach, Va.

Interviewer: How many American G.I.'s are buried at North Africa American Cemetery?

North Africa American Cemetery in Tunisia (Photo: ABMC).

Green: Approximately 2,841. They represent the entire North Africa campaign.

Interviewer: How many visitors make their way to the cemetery each year?

Green: Over last few years, we have averaged around 1,100 Americans and 4,000 from other countries. This has changed this year as we will have received twice that amount because of the cruise lines bringing Americans here on two separate occasions. We received more than 300 Americans per day.

Interviewer: How did you become involved in the ABMC?

Green: After I retired from the Marine Corps in 1990, I was living in Paris and looking for employment when I heard about the American Battle Monuments Commission. I had no idea until that day in July of 1997 that we had memorials overseas. I applied, and in November of 1997, was hired to start working at the Meuse Argonne American Cemetery in northeastern France. It is about 25 miles from Verdun, France.

About 8 months later, I was transferred to the Normandy American Cemetery and spent two years there. I was eventually promoted and transferred to American Cemetery in Suresnes, France, which is 5 miles west of the center of Paris. While there, I realized that we needed to remind the American people about our memorials and cemeteries abroad. In May 2002, I volunteered to come to Tunisia and have been here since.

Interviewer: Are there interesting landmarks or stories about the men and women buried at the cemetery?

Green: There are many. There is a monument to the HMS *Rhona*—which was sunk by a Nazi guided bomb on 26 Nov 1943 off the coast of Algeria. More than 1,000 soldiers lost their life in that sea disaster. Another monument is to the SS *Paul Hamilton*, which was en route to Sicily on 20 April 1944, when it was attacked and sunk and everyone on board was killed—that's 580 soldiers and sailors. The *Lady be Good*, a B-24 bomber, took off from an air base—never to be seen again until it was found in the desert in 1959 by some British oil workers. Between 1959 and 1960, recovery teams found the remains of the entire crew except for Vernon L. Moore whose name is on wall of the missing. Foy Draper, who earned the Olympic Gold Medal in the 1936 Berlin Olympics for running the 400-meter relay, is buried here. They are all heroes—every one of them. These are a few of the stories. You need to come and visit to hear the rest.

Interviewer: The cemetery also is on part of the Carthaginian and Roman city. Does any of it remain for visitors to see?

Green: 100 meters away from the cemetery is remains of the city, which is all around us.

Interviewer: What hidden, or unknown sites, are your favorites in the cemetery?

Green: Nothing is really hidden or unknown. However, the Osmun brothers are buried side by side. They were killed together on Christmas Eve 1942—and it is the most humbling reminder that brothers fought side by side and died in battle together.

Interviewer: Are there any interesting historical sites you would recommend tourists see in Carthage or Tunis?

Green: The list is long, but after visiting this American Memorial and Cemetery, there is the third largest Roman coliseum in the world along with all the ruins of Carthage. Hannibal was born here and his birth place is not far from the cemetery.

FACTS ABOUT THE NORTH AFRICA AMERICAN CEMETERY

Actual interments at North Africa American Cemetery began on 14 September 1948. On 1 July 1949, the cemetery was transferred to the authority of the American Battle Monuments Commission. Construction of the cemetery was completed in 1960 and the dedication ceremony was held on 21 July 1960. Gen. Dwight Eisenhower, Army Chief of Staff in 1946, was instrumental in the selection of the site in Carthage. Prior to 1948, burials had taken place in numerous temporary cemeteries throughout North Africa - seven in Tunisia, three in Algeria and Morocco, one in Teheran and at Lagos Field, Nigeria. Most of these temporary cemeteries were near the battlefields where the men fell. In 1948, the American Battle Monuments Commission offered the families the choice of repatriating the remains to the United States or permanent burial in Carthage.

One Medal of Honor recipient, Pvt. Nicholas Minue, Co. A, 6th Armored Infantry, 1st Armored Division, 28 April 1943, is interred at E-8-4. The remains of two other Medal of Honor recipients during Operation Torch, Pvt. Robert D. Booker, 133rd Infantry Regiment and Sgt. William Nelson, 60th Infantry Regiment, 9th Infantry Division were repatriated to the United States.

Four sets of brothers are buried here side by side. The Osmun brothers were killed in action on 24 December 1942. One brother of another pair was killed in Tunisia in 1942—his sibling died later in Brittany. The sibling's remains were eventually transferred from France to Carthage so that they could rest side by side.

Seven "unknowns" are interred at E-5-16. Two side-by-side graves at I-10-4 and I-10-5 hold the remains of four soldiers identified on a bronze plaque only as a group.

Four WACs are buried here. There are another 18 women memorialized on the Tablets of the Missing.

On 26 November 1943, the HMS *Rhona* carrying troops to Burma and India to construct airfields was struck by one of the first German glider bombs used against the Allies. There were 2,000 troops aboard, 195 sailors, 25 officers and 13 other personnel. Official records list a total of 1,050 Americans lost, which ranks very close to the losses on USS *Arizona* during the Pearl Harbor attack two years earlier. Interred at this cemetery are 31 who were on the HMS *Rhona* and 899 are memorialized on the Tablets of the Missing.

Capt. Foy Draper, 97th Bomb Squadron, 47th Bomb Group was a Gold Medalist in the 1936 Olympics in Berlin is buried here. Capt. Draper was on the 400-meter Relay Team with Jesse Owens, Roy Metcalf and James Wykoff. They set a blazing 39.8 second record that was not broken until 1956.

On 20 April 1944, the U.S. troop ship SS *Paul Hamilton,* carrying 471 U.S. Army Air Corps ordnance experts with their ordnance, blew up. A total of 580 men died in this tragic explosion; 520 are memorialized on the Tablets of the Missing.

CHARLES GROJEAN: SUBMARINER WHO HEADED THE NIMITZ FOUNDATION

One of the U.S. Navy's submarine leaders has been the driving force behind the Nimitz Foundation and National Museum of the Pacific War, retired Rear Adm. Charles Grojean, served as executive director of the Nimitz Foundation. The Foundation provides private financial support to expand and enhance the collections, exhibits, and programs of the National Museum of the Pacific War located in Fredericksburg, Texas. It also works to further research and document the historical record of the Pacific War.

We spoke with Adm. Grojean, before his death in 2008, about the mission of the foundation and museum. Grojean attended the Virginia Military Institute for a year before accepting an appointment to the U.S. Naval Academy. After graduation from the Academy in 1945, he served four years in the "black shoe" Navy before reporting to New London in November 1949 to begin training for his new career in the "silent service."

After earning his dolphins, Grojean served in both diesel/electric and nuclear submarines in the Atlantic and Pacific fleets. He later commanded *Angler*, *Barb* and *Thomas Edison*, as well as Submarine Flotilla Eight and Submarine Force Mediterranean (Naples), Italy.

Grojean, known as Chuck to his friends, entered the submarine service at that critical time when the nuclear submarine became a primary means by which the U.S. Navy projected American power on the high seas. Under the driving tutelage of the controversial Adm. Hyman G. Rickover, the U.S. Navy's and the world's first nuclear powered boat, USS *Nautilus*, slid down the ways in January 1954. It was quickly followed by a succession of nuclear boats, both hunter/killer/attack types designed to track Soviet submarines by utilizing our sophisticated electronic capability coupled with our knowledge of undersea thermal layers and the "boomers," boats deploying our Polaris ICBMs. The new undersea nuclear Navy in which Grojean served with such distinction had become a vital part of America's defense strategy during the Cold War with the Soviet Union.

Grojean's later shore assignments included duty at the Naval Academy, service on the staff of the Secretary of Defense, the Joint Chiefs of Staff and the Chief of Naval Operations, where he was director of political affairs.

Interviewer: What is your involvement with the Nimitz Foundation?

Grojean: I retired from active duty in June 1975. At that time, I was offered the position of superintendent of the Iranian Naval Academy, which was going to be located on the Caspian Sea. That fell through, so I moved to San Antonio and became a venture capitalist. Although that was a profitable venture with great people, I decided to become executive director of the Nimitz Foundation in July of 1990. It was a pro bono position, because at the time, the Foundation had no money. We have since come a long way. Our annual budget is now in the neighborhood of $500,000 a year. In addition, we have raised more than $6 million since 1994 to enlarge the museum and build The National Museum of the Pacific War. We have become quite an active group since our beginning. We are currently conducting a campaign to raise $12 million.

Admiral Chuck Grojean

Interviewer: That money will go toward a new facility?

Grojean: Approximately $9 million will go toward a new facility. We are hoping that the state of Texas will help. We are also hoping to raise $3 million from the private sector to go toward an endowment to assist with the operation of the museum.

Interviewer: How has the museum evolved over the years?

Grojean: There was a group of people in Fredericksburg who decided they would buy the old Nimitz Hotel, which had been owned by the grandfather of Adm. Nimitz. Their idea was to buy the hotel and create a museum named for Adm. Nimitz. When the group approached the admiral with the idea, he agreed that his name could go on the museum. But he requested that hotel building be restored to the condition it was when his grandfather owned it. In addition, he wanted the museum dedicated to all of the men and women who had served with him in the Pacific. So that building, which looks like a steamboat, was home for the museum. In 1994, we decided to go out and raise the money to build a larger facility that would better house our growing collection.

Interviewer: What are highlights of your collection?

Grojean: We have the best collection of Pacific war artifacts in the world. We even have the 80-foot, two-man Japanese submarine captured at Pearl Harbor. I have spoken with the pilot of the sub, Mr. Sakamaki, who became prisoner of war No. 1 after he was captured. Mr. Sakamaki returned to the United States in 1991 for the 50th anniversary of the attack on Pearl Harbor. He was concerned that someone may want to harm him if he came back, but I assured him that he would be safe, that we had put the war behind us and we were just looking for history. We had a terrific memorial symposium—more than 1,000 attended, including former President George Bush.

Interviewer: I noticed that you have an area specifically dedicated to President H.W. Bush's service.

Grojean: Yes, the main museum, which is the George Bush Gallery, is one we plan to expand by another 40,000 square feet. In there, we have the casing for what would have been the No. 3 or No. 4 atomic bomb dropped on Japan. In addition we have an area called the Pacific Combat Zone. We have five huge buildings. One building has a replica of the type of plane that President Bush flew on an aircraft carrier hanger deck. In the next building, we have a PT boat base–like those in the South Pacific in 1942. When you go through the buildings on a guided tour, there are mannequins on the exhibits that "talk" and so forth. We also have a World War II Quonset hut, where we have a medical exhibit. It is very moving, a lot of people come out of there in tears.

Interviewer: How did you acquire these pieces?

Grojean: A lot of the things like tanks and guns, the son of my predecessor here, Doug Hubbard, went over to New Guinea, New Britain and other places to get them back here. Many of them are actual pieces that were used in service and picked up right off the battlefields. Some are rusty. A lot of this work was done in the 1960s.

Interviewer: How many visitors does the museum have each year?

Grojean: Depending on the year, from 85,000 to 105,000.

Interviewer: What is the admission fee?

Grojean: The fee is now $5 for adults. Children are $3, but we can make special arrangements for school classes. We have 15,000 students come through each year.

Interviewer: You served for many years in the U.S. Navy, what are some of your best memories?

Grojean: I served for 30 years. I think the highlight was when I commanded a submarine force in the Mediterranean–my home base was Naples, Italy. Professionally, it was very rewarding. It was right in the middle of the Cold War. My subs and the Russian subs were always following each other, keeping track of what each of us was doing. I was there from 1968 to 1971.

Interviewer: What was your reason for choosing the submarine service?

Grojean: You got a command early, it's a smaller outfit and there is a lot of camaraderie. It was very exciting. I went into submarines before I was aware of the nuclear power aspect of it. Nuclear power was the best thing that happened to us. I had command of a diesel sub, then a nuclear attack sub, and then I went to a nuclear missile sub.

Interviewer: What was the name of the first nuclear boat you commanded?

Grojean: That was the *Barb*, named after the famous World War II submarine Barb, which was in the top three of sinking enemy tonnage.

Interviewer: I am sure that you had many exciting experiences. Does anything specifically come to mind about your service?

Grojean: There were always near misses, every time you read of a collision, a submerged collision, you think, 'there but for the grace of God go I.' There were many times things went wrong – tension was very high. You always figure that you'll get out of it, and we always did, but the *Thresher* didn't get out of hers, and the *Scorpion* didn't get out of hers; they both sank. The *Thresher* on a sea trial and the *Scorpion* en route back to its home port in Norfolk. But it never arrived. We put out a search team. But after several weeks of searching, we found the ship on the bottom of the ocean. It had imploded.

Interviewer: Did you know Admiral Rickover?

Grojean: Oh yes, very well. I served with him on his staff for a whole year and trained for nuclear—and I saw him every day. He was a very unique, dynamic, controversial individual. Every nuclear officer that went into the submarine force from the 1950s to the 1980s had to be approved by him. He was the czar of the submarine service. Unfortunately, his tactics drove a wedge between the submarine force at the time. Some guys who were not accepted were resentful, but they usually got over it.

Interviewer: Concerning the Pearl Harbor controversy, have you read *Scapegoats*, which defended Admiral Kimmel and General Short?

Grojean: I have not read it, but I have a file about two-feet-thick on the whole thing.

Interviewer: I suppose you have more information on it than the average Joe.

Grojean: In fact, Kimmel's son was my squadron commander when I commanded my first diesel submarine in Groton, Conn.

Interviewer: Was he highly thought of?

Grojean: Very highly. All of the Kimmels were well thought of.

Interviewer: What were you doing when you heard of the attack on Pearl Harbor?

Grojean: I was on my way back from church at Virginia Military Institute, passing Robert E. Lee's tomb, and another student told me about the attack. I thought it was bum dope. It was impossible for any battleship to ever be sunk.

Interviewer: Any other thoughts why someone should visit the museum?

Grojean: Your readers should know that the museum is now very large and everybody who goes through thinks it's great. We are only 73 miles from San Antonio. The drive from there is beautiful – and right through the Texas Hill country!

APPENDICES

1. How to Interview Veterans

Interviewing a combat veteran can be a very rewarding and educational experience. The following tips will allow you to have a respectful and informative interview to preserve a veteran's story for future generations.

Do your research. Be prepared! This may be the first and most important piece of advice I give to those interviewing a veteran. Learn about the veteran's military service, including the war or conflict they fought in, their rank, and their unit. Find out if the unit served during a specific campaign or battle. This will help you ask more informed questions and show the veteran that you are interested in their story.

Establish rapport. If you are a family member of a veteran, you probably have established a rapport with them. However, if you are interviewing a person you do not know, begin the interview by introducing yourself and expressing your appreciation for their service. It's important to build a connection by sharing common ground and expressing genuine interest in their story. I interviewed a

P-51 Mustang pilot who flew cover for the *Enola Gay* over Hiroshima. I asked him, did you know what that airplane was dropping and what effect it would have—or were you given a huge surprise when that bomb went off? He said, with a smile, "We knew."

Ask permission. Ask permission to record a conversation and to take notes.

Be patient and kind. Be understanding and patient with veterans, particularly older vets. Their memories have become faded. Sometimes they are emotional as they experienced very traumatic wartime experiences. Nearly all veterans of World War II who saw combat have harrowing stories. Treat the veteran with respect and kindness. If they need a break, let them have one.

Listen. The veteran who is willing to talk with you (many will not discuss their participation in war) is someone to be listened to very carefully. Maintain eye contact, encourage them and be engaged. Don't interrupt them and respect their boundaries as some topics may be off-limits.

Ask questions that invite storytelling. I always ask them what they did before and after the war. I always ask veterans what they liked about military service—and what they didn't like. Plan questions ahead of the interview and don't be surprised if the veteran doesn't answer the way you think they should. Don't make assumptions.

Here are some specific questions you could ask the veteran:

- What was your motivation for joining the military?
- What was your experience like during basic training?
- What was your first deployment like?

- What was the most difficult thing you experienced during your time in combat?
- What was the most rewarding thing you experienced during your time in combat?
- How did your military service change you?
- What advice would you give to young people who are considering joining the military?

These are just a few examples of questions to ask a veteran. It's important, based on your previous research, to tailor your questions to the veteran's individual story. For example, when interviewing Bill Geiger, who was an American serving in the Royal Air Force, it was important to ask about "The Great Escape." Geiger was a prisoner of war at Stalag Luft III at the same time the famous escape occurred.

As you have read in a few of the interviews in this book, once the veteran has become comfortable with a question, you have the opportunity to their opinion on a specific incident or battle. One question I asked veterans, of both the Pacific and European theatres, is whether they agreed with the decision to drop the atomic bomb. Another, which I asked Donald Beaty, was "do you think the [USS *Indianapolis*] captain was a scapegoat" for the incident?

The National World War II Museum, through its Oral History Program, has a number of solid questions to ask World War II veterans. They can be found at:

https://www.nationalww2museum.org/oral-history-resources.

Thank the veteran. Thank them for their service and their time.

Follow up with the veteran or their caregiver. Thank the veteran and give them your contact information. Also tell them what you are going to do with the interview. World War II Marine veteran Don Whipple called me back when he forgot an important part of his experiences on Iwo Jima He said it was important for future generations to understand what happened there. I gave him my phone number so he could reach out at any time. However, respect the veteran's wishes if they want to be anonymous.

Some additional suggestions:

Record the interview. In our next section, we will tell you where to upload/send a veteran's interview—whether that be a voice recording or a video. These are extremely valuable historic documents. Once the World War II generation is gone, these will be priceless records of men and women who experienced Pearl Harbor, D-Day, the atomic bomb drops, the Holocaust, and other key events.

Transcribe the interview. Not only will it be easier to share a veteran's story, but you can go back to reference notes for future interviews.

Get the veteran's permission to share their story. This is especially important if you plan to publish the interview or use it in any other way.

What to do if your veteran is not available? Sometimes, it is not possible to do an interview in person, by telephone, or by email. This could be for many reasons including the health of the veteran, availability of a relative or caregiver to give access to the subject, or other reasons. You can write a story about the veteran without an in-person interview. You can do your own research by looking at the veteran's or family member's social media accounts (Facebook, X, Instagram, etc.). Often, the family will allow you to use photos from social media accounts for your story. Always ask for permission. In addition, you can add background information to a story from newspaper, magazine, and online articles. Always mention where the story was published and who wrote it. Examples of this type of writing can be viewed in our story about Albert Consigli. A hybrid version of this type of story was the Charles Norman Shay interview, which relied heavily on previously published material.

2. Preserving Your Veteran's Story

After you are granted permission by the veteran, preserving and archiving your stories and interviews are important documents for future generations. Don't forget to add the veteran's name, rank, and service branch when you archive an interview. Also include the date and location of the interview, a brief description, and a transcript or recording.

Here are some suggested places to archive your interviews:

Library of Congress Veteran's History Project

The United States Congress, in 2000, created the Veterans History Project (VHP) as part of the American Folklife Center at the Library of Congress. VHP collects, preserves and makes accessible the personal accounts of American war veterans so that future generations may hear from them and gain a better perspective about the realities of war.

As part of the program, veterans can record 30-minute or longer conversations/interviews about their military experiences. In addition, veterans can collate 20 or more pages of memoirs, diaries/journals; provide 10 or more photographs and letters and submit them to VHP for the project.

Because all submissions become the property of the Library of Congress, VHP suggests that copies be made of each document. It is suggested that the friends and families of the veteran assist in the submission of documents.

VHP says that educators, researchers and scholars rely on the collection to supplement historical texts and support research projects. After submission, family members, friends and other loved ones will treasure the memories that are permanently preserved for future generations at the Library of Congress, the VHP says.

VHP field kits are available that can be used by veterans, family members and even the Boy Scouts to record wartime stories. These can be downloaded and printed from the VHP website at: www.loc.gov/vets.

Contact:

(800) 371-5848

www.loc.gov/vets.

The National World War II Museum.

According to its website, the National WWII Museum "strives to preserve the legacy and lessons of World War II through the stories of those who experienced the war." The museum accepts oral histories and memoirs that people have conducted or printed themselves.

Mail to:
The National WWII Museum
Attn: Museum Library
945 Magazine Street
New Orleans, LA 70130

The museum also offers an Oral History Program with background information, tips and common questions to ask both veterans and those who lived on the home front.

Contact:

oral.histories@nationalww2museum.org
(504) 528-1944, ext 516
www.nationalww2museum.org/oral-history-resources.

National Museum of the U.S. Army's Registry of the American Soldier

The National Museum of the U.S. Army's Registry of the American Soldier provides an opportunity for veterans to have his or her name and service history, in war and peace, placed on record at the museum. Former and currently serving soldiers also can register the names of family members if they currently serve or have served in the U.S. Army.

The museum says this is a unique way to document you or your family's service. The registrant can control the content and can change or add information at any time. This includes the participant's rank, hometown and service activity.

Contact:

The National Museum of the U.S. Army
(800) 506-2672
www.armyhistory.org.

U.S. Army Heritage and Education Center

The U.S. Army Heritage and Education Center (USAHEC) is a great place for historians, researchers and soldiers to get access about the men and women who shaped the service. Thousands of veterans have shared their stories with USAHEC.

Contact:

USAHEC
(717) 245-3972
usarmy.carlisle.awc.mbx.ahec-ves@mail.mil.

Other ways to preserve and archive interviews:

- **At your local historical society or museum**. These historical societies and museums often have archives that preserve veterans' stories, particularly from local service people.
- **Universities or nonprofit research institutions**. Universities and research institutions have archives that focus on graduates or even different wars. Often they are managed by the libraries, an on-campus museum, or the history department.
- **Veterans' organizations**. Such organizations as the American Legion, Veterans of Foreign Wars and others have programs dedicated to their individual post's members.
- **Your library**. Many local libraries have oral history programs that detail the service of area residents and units.

- **A website, blog or YouTube channel**. You can place the veteran's interview, video and other content on your own website. There are many online resources on how to purchase a domain name, get a WordPress or other website created, and find a hosting service. After that, you can link the website to your social media (Facebook, X, and others). You also can create a YouTube channel by uploading your video and sharing it.

3. Suggested Reading

Here are a few very good books that we have culled from the large number published about World War II. They are currently available through Amazon.com and other outlets. These books will help you understand where your veteran interview subject served and what were the overall objectives and results. The books will also help you prepare your interview questions.

Histories

Ambrose, Stephen E. *D-Day, June 6, 1944: The Climactic Battle of World War II* (1994).

- *Pegasus Bridge: June 6, 1944* (1985).

- *Band of Brothers: E Company, 506th Regiment, 101st Airborne from Normandy to Hitler's Eagle Nest* (1992).

Beevor, Antony. *D-Day: The Battle for Normandy* (2003).

D'Este, Carlo. *Decision in Normandy* (1983).

The Battle of Normandy: Operations Manual (2013).

Hastings, Max. *Overlord: Day-Day and the Battle for Normandy* (1984).

Keegan, John. *Six Armies in Normandy* (1982).

Kershaw, Alex. *The Liberator: One World War II Soldier's 500-Day Odyssey from the Beaches of Sicily to the Gates of Dachau* (2012).

Lewis, Adrian R. *Omaha Beach: A Flawed Victory* (2001).

Miller, Donald L. *Masters of the Air: America's Bomber Boys Who Fought the Air War Against Nazi Germany* (2006).

Morison, Samuel Eliot, *History of United States Naval Operations in World War II* (1947-1962). *The Two-Ocean War: A Short History of the United States Navy in the Second World War* (1963).

Nightengale, Keith. *The Human Face of D-Day: Walking the Battlefields of Normandy* (2023).

Ryan, Cornelius. *The Longest Day, June 6, 1944* (1959).

Spector, Ronald H. *Eagle Against the Sun: The American War with Japan* (1985).

Stanton, Doug. *In Harm's Way: The Sinking of the USS Indianapolis and the Extraordinary Story of Its Survivors* (2001).

Wieviorka, Olivier. *Normandy: The Landings to the Liberation of Paris* (2008).

Memoirs

Bradley, Omar N. *A General's Life; An Autobiography* (1984).

Burgette, Donald R. *Currahee!* (1967).

Cawthon, Charles R. *Other Clay. A Remembrance of World War II Infantry* (1990).

Eisenhower, Dwight D. *Crusade in Europe* (1948).

Gavin, James M. *On To Berlin* (1978).

Sledge, Eugene B. *With the Old Breed: At Peleliu and Okinawa* (1981).

4. Movies and Television Series

Movies and television shows about war and military service often come with political and/or Hollywood points of view and should be taken with a grain of salt. However, they are a great place to start to put your veteran interview subject's service into perspective. What better to do that than watching the opening scene of *Saving Private Ryan*. However, even that well-made movie was panned by D-Day veterans, who told me in an interview in 2001, that while the landing scenes were very realistic, the movie evolved into just another Hollywood yarn.

Here's a small sample of good World War II

Band of Brothers (2001).

Battleground (1949).

The Best Years of Our Lives (1946).

A Bridge Too Far (1977).

Flags of Our Fathers (2006).

The Great Escape (1963).

The Longest Day (1962).

Letters from Iwo Jima (2006).

Masters of the Air (2024).

Midway (1976).

The Pacific (2010).

Patton (1970).

Sands of Iwo Jima (1949).

Saving Private Ryan (1998).

Schindler's List (1993).

The Story of G.I. Joe (1949).

Tora! Tora! Tora! (1970).

ABOUT THE AUTHOR

Kevin Dennehy

Kevin Dennehy has been a journalist for more than thirty years, writing for daily newspapers and magazines. Dennehy is the author of the bestselling *The D-Day Visitor's Handbook.* A retired Army National Guard colonel, he is a Special Forces combat veteran of Afghanistan and Iraq.

INDEX

ALSO FROM AUTHOR KEVIN DENNEHY AND AVAILABLE ON AMAZON.COM:

Made in United States
Troutdale, OR
11/03/2026

56844459R00082